Judith Smait

A Guide to
Marine Coastal Plankton and Marine Invertebrate Larvae

Second Edition

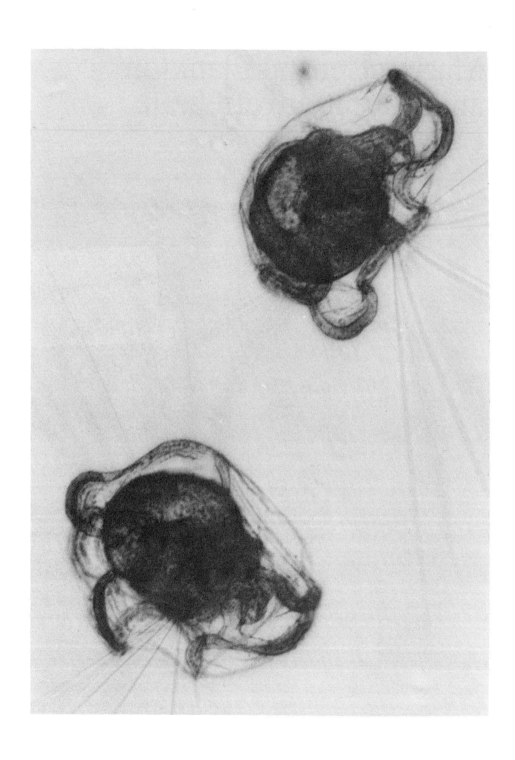

A Guide to
Marine Coastal Plankton and Marine Invertebrate Larvae

Second Edition

DeBoyd L. Smith
 and
Kevin B. Johnson

KENDALL/HUNT PUBLISHING COMPANY
4050 Westmark Drive   Dubuque, Iowa 52002

Front cover: Brachiolaria larva of the sea star *Pycnopodia helianthoides*. Photo reprinted by permission of Richard B. Emlet.

Back cover: Foraminiferan *Globigerina* sp. Photo courtesy of Cynthia L. Roye.

Copyright © 1977 by DeBoyd L. Smith
Copyright © 1996 by DeBoyd L. Smith and Kevin B. Johnson

Library of Congress Catalog Card Number: 96-76343

ISBN 0-7872-2113-9

All rights reserved. No part of this publication may be reproduced, stored in a retrieval system, or transmitted, in any form or by any means, electronic, mechanical, photocopying, recording, or otherwise, without the prior written permission of the copyright owner.

Printed in the United States of America
10 9 8 7 6 5 4 3 2 1

Contents

List of Plates	*vii*
Foreword	*xi*
Preface	*xiii*
Acknowledgments	*xv*

Introduction — 1

1 Plankton Collection, Observation & Culture — **5**
    Plankton Collection — 5
    Plankton Observation — 11
    Plankton Culture — 13

2 Quick Flip Reference — **15**
    Quick Flip Reference Plates — 16

3 Planktonic Protistans — **25**
    Diatoms — 29
    Ciliates — 43
    Foraminiferans — 46
    Actinopods — 46
    Flagellates — 49

4 Zooplankton Ecology — **55**
    Specialization — 55
    Motility — 56
    Feeding — 56
    Reproduction — 57
    Development — 58
    Taxonomy — 61

5	**Zooplankton Identification**	**63**
	Phylum Porifera	63
	Phylum Cnidaria	63
	Phylum Ctenophora	83
	Phylum Nemertea	85
	Phylum Platyhelminthes	89
	Phylum Nematoda	93
	Phylum Rotifera	93
	Phylum Sipuncula and Phylum Echiura	93
	Phylum Annelida	97
	Phylum Arthropoda	111
	Phylum Mollusca	153
	The Deuterostome Phyla	166
	The Lophophorates	167
	Phylum Ectoprocta (Bryozoa)	167
	Phylum Phoronida	169
	Phylum Brachiopoda	171
	Phylum Chaetognatha	173
	Phylum Echinodermata	175
	Phylum Hemichordata	183
	Phylum Chordata	185

Glossary	197
References	203
Index	215

List of Plates

1. Quick Flip Reference. Diatoms, Foraminiferans & Actinopods — 16
2. Quick Flip Reference. Ciliates, Dinoflagellates, Cnidarians & Misc. — 17
3. Quick Flip Reference. Ctenophores, Pilidia, Vermiformes & Eggs — 18
4. Quick Flip Reference. *Polygordius* (Polychaeta) & Crustaceans — 19
5. Quick Flip Reference. Various Crustaceans — 20
6. Quick Flip Reference. Crustaceans and Molluscs — 21
7. Quick Flip Reference. Molluscs and Deuterostomes — 22
8. Quick Flip Reference. Deuterostomes — 23

9. Diatoms, Class Coscinodiscophyceae — 30
10. Diatoms, Class Coscinodiscophyceae — 31
11. Diatoms, Class Coscinodiscophyceae — 32
12. Diatoms, Class Coscinodiscophyceae — 33
13. Diatoms, Class Coscinodiscophyceae — 34
14. Diatoms, Class Coscinodiscophyceae — 35
15. Diatoms, Class Coscinodiscophyceae — 36
16. Diatoms, Class Coscinodiscophyceae — 37
17. Diatoms, Class Coscinodiscophyceae — 38
18. Diatoms, Class Fragilariophyceae — 39
19. Diatoms, Class Fragilariophyceae — 40
20. Diatoms, Class Fragilariophyceae — 41
21. Diatoms, Class Bacillariophyceae — 42

22. Ciliophora (Ciliates) — 44
23. Ciliophora (Ciliates): Tintinnids and loricae — 45
24. Granuloreticulosa (foraminiferans) — 47
25. Superclass Actinopoda — 48
26. Representative Marine Flagellates — 50
27. Dinoflagellates (Phylum Dinoflagellata) — 51
28. Dinoflagellates (Phylum Dinoflagellata) — 52

29. Phylum Cnidaria: larvae, ephyrae and postlarval juveniles — 64
30. Phylum Cnidaria: Class Scyphozoa — 66
31. Phylum Cnidaria: Hydromedusae and Trachymedusae — 70
32. Phylum Cnidaria: Leptomedusae — 71
33. Phylum Cnidaria: Anthomedusae — 72
34. Phylum Cnidaria: Anthomedusae — 73
35. Phylum Cnidaria: Anthomedusae — 74
36. Phylum Cnidaria: Hydroid Polyps — 76
37. Phylum Cnidaria: Hydroid Polyps — 77
38. Phylum Cnidaria: Orders Chondrophora & Siphonophora — 79
39. Phylum Cnidaria: Order Siphonophora — 80

List of Plates

40.	Phylum Cnidaria: Order Siphonophora	81
41.	Phylum Cnidaria: Order Siphonophora	82
42.	Phylum Ctenophora (Comb Jellies)	84
43.	Phylum Nemertea: Unidentified Pilidium Larvae	86
44.	Phylum Nemertea: Pilidium Larvae and Juveniles	87
45.	Phylum Nemertea: Unidentified Planktonic Nemerteans	88
46.	Phylum Platyhelminthes: Turbellarian Flatworms, Larvae & Juveniles	90
47.	Phylum Platyhelminthes: Trematodes and Turbellarian Flatworms	91
48.	Phylum Nematoda and Phylum Rotifera	94
49.	Phylum Sipuncula: Trochophore and Pelagosphaera Larvae	95
50.	Phyla Sipuncula and Echiura: Unidentified Trochophore Larvae	96
51.	Phylum Annelida: Polychaetes, Early Trochophore Larvae	100
52.	Phylum Annelida: Polychaetes (Families Syllidae and Tomopteridae)	101
53.	Phylum Annelida: Polychaetes (Family Spionidae)	102
54.	Phylum Annelida: Polychaetes (Families Nereidae and Goniadidae)	103
55.	Phylum Annelida: Polychaetes (Representatives of Four Families)	105
56.	Phylum Annelida: Polychaetes (Representatives of Three Families)	106
57.	Phylum Annelida: Polychaetes (Representatives of Four Families)	107
58.	Phylum Annelida: Hirudineans and Four Polychaete Families	108
59.	Phylum Arthropoda: Classes Pycnogonida and Arachnida	112
60.	Phylum Arthropoda: Crustacea: Nauplius Larvae	114
61.	Phylum Arthropoda: Crustacea: Barnacle larvae and molts	116
62.	Phylum Arthropoda: Crustacea: Ostracods	117
63.	Phylum Arthropoda: Crustacea: Ostracods and Cladocerans	118
64.	Phylum Arthropoda: Crustacea: Common orders of pelagic copepods	121
65.	Phylum Arthropoda: Crustacea: Copepoda developmental stages	122
66.	Phylum Arthropoda: Crustacea: Calanoid copepods	123
67.	Phylum Arthropoda: Crustacea: Calanoid copepods	124
68.	Phylum Arthropoda: Crustacea: Cyclopoid copepods	125
69.	Phylum Arthropoda: Crustacea: Harpacticoid copepods	126
70.	Phylum Arthropoda: Crustacea: Parasitic copepods	127
71.	Phylum Arthropoda: Crustacea: Parasitic copepods	128
72.	Phylum Arthropoda: Crustacea: Mysid Shrimps	130
73.	Phylum Arthropoda: Crustacea: Mysid Shrimps	131
74.	Phylum Arthropoda: Crustacea: Cumaceans, Tanaids & Stomatopods	133
75.	Phylum Arthropoda: Crustacea: Isopods	134
76.	Phylum Arthropoda: Crustacea: Isopods and Gammarid Amphipods	135
77.	Phylum Arthropoda: Crustacea: Gammarid and Hyperiid Amphipods	137
78.	Phylum Arthropoda: Crustacea: Caprellid Amphipods	138
79.	Phylum Arthropoda: Crustacea: Euphausid developmental stages	140
80.	Phylum Arthropoda: Crustacea: Caridean Shrimp Larvae	142
81.	Phylum Arthropoda: Crustacea: Caridean Shrimps	143
82.	Phylum Arthropoda: Crustacea: Caridean Shrimps	144

List of Plates

83. Phylum Arthropoda: Crustacea: *Pagurus* larval stages	**146**
84. Phylum Arthropoda: Crustacea: Pagurid zoea larvae	**147**
85. Phylum Arthropoda: Crustacea: Anomura and Brachyura zoea larvae	**148**
86. Phylum Arthropoda: Crustacea: Brachyura zoea larvae	**149**
87. Phylum Arthropoda: Crustacea: Anomura and Brachyura megalopae	**150**
88. Phylum Arthropoda: Crustacea: Brachyura megalopae	**151**
89. Phylum Mollusca: Chiton developmental stages	**154**
90. Phylum Mollusca: Gastropod egg cases and veligers	**156**
91. Phylum Mollusca: Gastropod veligers	**157**
92. Phylum Mollusca: Gastropod veligers	**159**
93. Phylum Mollusca: Pelagic Gastropods	**160**
94. Phylum Mollusca: Nudibranch larvae and juveniles	**162**
95. Phylum Mollusca: Bivalve larvae and juveniles	**164**
96. Phylum Mollusca: Cephalopod juveniles	**165**
97. Phylum Bryozoa: Larvae and adult colony fragments	**168**
98. Phylum Phoronida: Actinotroch larvae, metamorphosis and juveniles	**170**
99. Phylum Brachiopoda (larvae) and Phylum Chaetognatha (adults)	**172**
100. Phylum Echinodermata: Developmental stages in sea stars	**177**
101. Phylum Echinodermata: Developmental stages in sea stars	**178**
102. Phylum Echinodermata: Developmental stages in sea cucumbers	**179**
103. Phylum Echinodermata: Echinoid larvae and juveniles	**181**
104. Phylum Echinodermata: Developmental stages in brittle stars	**182**
105. Phylum Hemichordata and Phylum Chordata (ascidian adults)	**186**
106. Phylum Chordata: Subphylum Urochordata: Ascidian Tadpoles	**188**
107. Phylum Chordata: Subphylum Urochordata: Appendicularians	**189**
108. Phylum Chordata: Subphylum Urochordata: Salps	**191**
109. Phylum Chordata: Subphylum Vertebrata: Larval Fishes	**193**
110. Phylum Chordata: Subphylum Vertebrata: Fish Eggs and Larvae	**194**
111. Phylum Chordata: Subphylum Vertebrata: Larval Fishes	**195**
112. Phylum Chordata: Subphylum Vertebrata: Fishes and Larval Fishes	**196**

Foreword

Dip your hand into the ocean and pick up a few drops of water in your cupped hand. This colorless, transparent liquid, so lacking in interest to the naked eye, if placed under the microscope will reveal a population of minute plants and animals. They live, grow, reproduce, and die in an endless cycle that makes the productivity of the oceans comparable to that of good farm land in the Sacramento San Joaquin Valley. The importance of these apparently insignificant organisms can scarcely be over-emphasized for we have recently come to realize that the continued existence of human life on this planet may actually depend upon the continuation of their biological cycle.

In addition to their importance to our well-being, these planktonic organisms claim our interest because they are strange, beautiful, grotesque, and have adapted to their environment in manifold marvelous ways. It is an inert student indeed who cannot sit fascinated over the microscope for hours simply watching and admiring them. But no real student is content to watch. They want to know.

As knowledge of the plankton has accumulated, it has been presented in many thousands of specialized papers appearing in many different journals published in various languages. Because of the problem of finding and accumulating the pertinent literature, it has been most difficult for the beginning student to learn what kind of an organism they were observing. What slow and painful steps need they take if they want to know? To what degree will their difficulties dull their interest?

The present work is an attempt to alleviate the difficulties. It is a pioneering synthesis of our knowledge concerning the marine plankton of central California. Since pictures are much more effective than words in describing an object, the simple line drawings should enable the student to determine, without too much difficulty, what kind of an organism they are observing. Here is a passport to the joy of discovery.

Carmel, California, November 1971   Rolf L. Bolin

Note: The passing of Dr. Rolf Bolin, 1974, Professor Emeritus, Stanford University, left a void in the field of marine studies. His most appropriate remarks as a foreword to the first guide are a reminder of enthusiasm in teaching and of the stimulus of a contagious delight in new discovery. In dedication to Dr. Bolin, this volume is truly intended as a "passport to the joy of discovery".

D.L.S.

Preface

This second edition of *A Guide to Marine Coastal Plankton and Marine Invertebrate Larvae*, updated and reorganized from the first edition, is an expanded version of Smith's pocket handbook *A Guide to Marine Coastal Plankton*, published in 1971. This guide is designed as an introduction to marine plankton, or a quick reference for researchers needing general identifications.

Many marine organisms with complex life cycles spend all or a period of their lives in the plankton. For this reason, we have included in this guide some information about larval stages and life histories. This information, coupled with basic sampling methods, references and drawings to aid in identification, make this guide especially valuable for teaching. In some cases, demersal zooplankton and partial organisms are identified as these are commonly found in coastal plankton tows.

Most plankton illustrated herein were collected in the California bays of Monterey, Bodega and Tomales. Plankton tows were hauled vertically, as there is evidence that certain planktonic species given the opportunity (i.e. in the absence of thorough mixing of the water column) may be vertically stratified. Vertical plankton tows may give a better representation of the diversity of marine plankton, including invertebrate larvae.

While some plankton seem to be omnipresent, many are sporadic temporally and spatially. You may come across forms, and you certainly will come across species, not described in this volume. Do not be dismayed. Undescribed plankton are only evidence of the astounding diversity of marine plankton. Many of the genera are world-wide in distribution.

Plankton samples for classroom study are readily obtained with minimum expenditure of time and equipment. Some short instructions for the collection and maintainance of plankton are given in this volume. Those far from the coast and unable to make a field trip for collection may obtain live or preserved samples from east or west coast suppliers.

While this guide is introductory in nature, there is a great need for a comprehensive guide to Pacific and Atlantic coastal plankton. Many forms are well known to the specialists of the various plankton taxa, but no one has made an attempt to provide an integrated work. This guide is not intended as a substitute for such a definitive study and the compilation of a comprehensive work is encouraged.

The book is intended as a practical guide to common marine plankton forms, for use in the classroom and laboratory. Because few biology textbooks include an emphasis on planktonic and larval forms, this book is also recommended as a supplement to textbooks used in courses of general biology, marine biology, oceanography, zoology, and invertebrate zoology.

Most illustrations in this volume were drawn from photomicrographs of living plankton taken by Professor Smith over a seven year period of plankton study. Most of these illustrations were drawn by Floy M. Zitton at Arcata, California during the Summer and Fall of 1973. Additional illustrations were by

Sue Macias when she was a student at West Valley College. New drawings, adaptations and line-drawing alterations from the first edition of *A Guide to Marine Coastal Plankton and Marine Invertebrate Larvae* were prepared by Kevin B. Johnson.

The first edition of this book has found widespread use in all fifty states and several foreign countries. Because many families and genera are widely distributed, this book has been found to be valuable in the waters of the Atlantic Ocean, the Carribean and the Gulf of Mexico in addition to the Pacific coast of North America. It is in use at all levels of education, including elementary schools, universities, research laboratories and governmental agencies. It is hoped that this book will encourage the study of marine plankton in the classroom and laboratory and contribute to a further understanding of the complex world of marine plankton. This volume aids in the study of the ecology of plankton assemblages and the developmental patterns of invertebrate life cycles.

Acknowledgments for the First Edition

Many individuals have been an influence in the preparation of this book. F.C. Van Buren first encouraged the author to return to school after World War II, Mel Griffin first stirred an interest in biology. Drs. Burch, Arvey, Shipley, and Hardy at Long Beach State College added greater stimuli, as did Henry Childs. Marine interests came under a N.S.F. grant to Hopkins Marine Station where such leaders as Drs. Bolin, Kinne, Hollenberg, the Abbotts, and others shaped further interests during five summers and an academic year. Plankton became an interest during early teaching days of West Valley Community College and several years of collecting, photographing, identification, etc., led to recognition of the need of a basic guide to plankton. A sabbatical leave, with use of the facilities at Pacific Marine Station, Dillon Beach, courtesy of University of the Pacific, during the school year 1971-72 brought the first *Guide to Marine Coastal Plankton*. Dr. James Blake, Dr. Wm. Gladfelter, and others were most helpful in that preparation. The added help of many who have used the first volume has been welcome in this publication. In particular Maxwell Eldridge has been of help with the fish larvae. Floy Zitton and Sue Macias in the preparation of the artwork, and Betty Bradler with the typing. As with the first edition, an expression of gratitude to my wife, Eileen, and our sons and daughter for their continued interest and encouragement should be noted as of prime importance. For all these and for my students, **past, present,** and future, I am grateful.

DeBoyd L. Smith

Acknowledgments for the Second Edition

As with the first edition of *A Guide to Marine Coastal Plankton and Marine Invertebrate Larvae*, many people have contributed to the author's ability to complete this project and understand the material. Professor DeBoyd L. Smith afforded me a welcome opportunity to work on this second edition. I was first introduced to these fascinating organisms by Dr. Lee F. Braithwaite of Brigham Young University, whose patience and enthusiasm helped me gain an appreciation of marine plankton. A special thanks to Dr. Greta Fryxell, professor emeritus of Texas A&M University, and Dr. Richard Strathmann and Megumi Strathmann of the University of Washington for sharing their expertise and giving input on this book. Much appreciation to Barbara Butler for performing above and beyond the call of duty. I would like to thank my advisor Dr. Alan Shanks and my advisory committee for allowing me to complete this book during my graduate tenure. This endeavor could not have been completed without the generosity and support of Jim and Bonnie Thompson and Royle and Sue Johnson. Finally, I would like to thank Colleen and Bethany Johnson for their invaluable support.

Kevin B. Johnson

Introduction

A miniature, breathtaking world of beauty is hidden beneath the surface of the ocean. Microscopic plants and animals of the plankton, with myriad contrast in form, function and design, will quickly draw your fascination and captivate your interest. Diatoms, other protistes and bacteria comprise the base of the planktonic food web. Minute jellyfishes, comb-jellies, various worm phyla, crustaceans and snails all play a role in the interactions of the plankton assemblage. Tadpole-like tunicates and newly hatched fishes also belong to this vast marine community. Each adds to the variety of planktonic life. Many are early developmental forms (e.g., eggs, blastulae, gastrulae, and larvae) of **benthic** marine invertebrates and are often radically different in appearance from the nonplanktonic adults. Larvae come in variable, sizes, shapes, colors and modes of locomotion and feeding. Most are less than 1 millimeter in size. Many are measured in terms of microns, visible only with the aid of a microscope. Observing these miniature plants and animals, we find a delightfully surprising community unique from any in our experience. Here is a new and exciting world to investigate.

There is now growing awareness of the importance of the oceans and the vast array of life they support. Research in many branches of oceanography is probing the unknown realms of the marine world. The public is increasingly aware of the upcoming importance of marine aquaculture in providing food for the bulging population of the world. Marine food and mineral resources promise substantial future exploration and exploitation in an effort to provide needed commodities. Public and private firms are becoming aware of these critical needs. Educational institutions, teachers, and students are looking more and more to a better understanding of this new world. Though much has been learned of our oceans' resources, many of their secrets and worth are still unknown.

There is a rapidly expanding interest in better understanding the marine environment and the role populations and individuals play in shaping complex marine communities. Those organisms that live on or attached to the sea floor, or benthos, are said to be benthic. Many familiar algae and invertebrates of the tidepools are benthic. Organisms that swim freely and determine their own regional distribution by doing so are considered **nektonic**. Organisms that drift in the ocean and are not strong enough swimmers to determine their own regional distribution are said to be **planktonic** (plankton from the Greek word for "wandering"). Both planktonic and nektonic forms live in the water column (somewhere between the sea floor and the surface of the water. In general, the nekton are the larger organisms while planktonic forms are smaller. In our coastal waters most plankton are microscopic, yet many are visible to the naked eye. The animals in the plankton are called **zooplankton** and the plant forms and many planktonic protistans are referred to as **phytoplankton**.

The precise definition of plankton has always been vague because many of the smaller forms are capable swimmers, some extremely fast. For some, the true plankton are the undisturbed, blue-water drifters of deeper, offshore waters. For others, plankton is whatever is caught in the "haul" of the plankton net (a fine-mesh silk or nylon net hauled upward from the ocean's depths or towed behind a boat). Plankton, captured in a net, is collected in a bottle attached to the lower, closed end of the net.

Many plankton are found in the blue waters of the open ocean, the **pelagic** zone. Other plankton are found in the shallower coastal waters, the **neritic** zone. **Meroplankton** are organisms that spend only part of their life-cycles in the plankton, while **holoplankton** are planktonic their entire lives. Most meroplankton tend to be neritic, because there comes a time when they must settle out of the plankton and into tidepools or other near-shore environments. Holoplankton are found everywhere, and are found more often than meroplankton in pelagic waters. Plankton are sometimes categorized according to the size class into which they fall. Classification systems are numerous. We present a system from Wimpenny (1966) which breaks down and labels plankton size classes as follows:

Picoplankton	0-2 μm
Ultraplankton	2-5 μm
Nannoplankton	5-60 μm
Microplankton	60-500 μm
Mesoplankton	500-1,000 μm (0.5-1.0 mm)
Macroplankton	1,000-10,000 μm (1.0 mm-1.0 cm)
Megaloplankton	greater than 10,000 μm (1.0 cm)

Introduction

Plankton can range in size from microns to meters. Among the organisms greater than 60 μm (microplankton to megaloplankton) are many of the herbivores (grazers) and carnivores (predators) of the plankton, each playing a part in the complex marine food web. The microplankton contain most of the diatoms and early larval forms of meroplankton. The smaller groups consist of flagellates, both **autotrophic** and **heterotrophic**, and the increasingly recognized marine bacteria. The selection of the plankton net's mesh size will influence the composition of the haul. The distribution of plankton in the ocean is dependent upon many variable factors including the intensity of light, time of day or night, salinity, temperature, turbidity, currents and tides, nutrients, seasonal reproductive cycles, and predators. The presence of sunlight in the upper 100-200 meters of the ocean, the **euphotic zone**, provides energy for the photosynthetic processes of the ocean's primary producers, including diatoms. Below the light, in the **aphotic zone**, many plankton can also be found. However, these cannot photosynthesize unless they have a mechanism for occasionally rising into the euphotic zone.

The illustrations in this guide were drawn from slides of plankton photographed live during a continuing microscopic study of coastal plankton samples. Many of the forms and genera represented are worldwide in distribution.

Here in the waters of the ocean is an unseen world of unique beauty, pattern, form and design. This miniature world is that of the plankton, tiny representatives of the kingdoms of life: the zooplankton, the glass-housed diatoms and the many other protistans. As with other ecosystems, in the plankton there are predators, prey, symbioses and communities. Recent oceanographic research indicates that **trophic** dynamics in planktonic systems may be much more complex than previously assumed. The relationship of predators and grazers with their planktonic prey more closely resembles a food web than a food chain. Protistans and bacteria are probably of paramount importance for cycling nutrients in planktonic systems. There is evidence that much of the available nutrients in planktonic systems are used by picoplankton in a cycle of grazing and excretion termed the **microbial food web**. Most of

Introduction

the more notorious ocean creatures (fishes, whales, seabirds, etc.) depend ultimately upon the plankton responsible for **primary production** at the base of the ocean's food web.

This guide is geared primarily toward the study of zooplankton. However, phytoplankters, including the common diatoms and dinoflagellates, are pervasive and some reference should be available for identifying the nonanimals. Besides playing an integral part in the ecology of zooplankton, planktonic protistans are an interesting study unto themselves. Phytoplankton come in a dazzling range of forms, including discs, rods, chains and spines. Many phytoplankton exhibit motility. Some mobile dinoflagellates are heterotrophic and roam in search of prey.

Observing these miniature organisms while they are living is a delightful change from the colorless, preserved specimens so often found in the biological laboratory. The investigator can watch eggs divide, or watch larvae and embryos change and grow. One can study how the organisms react to the elements of their environment. It is worthwhile and interesting to compare appendages between taxa, observe working gills, or see how tentacles, antennae and cilia have been adapted for feeding, locomotion or respiration. This is an exciting new world to investigate. This guide will help identify what is discovered. The world of the plankton is yours to explore.

Chapter 1
Plankton Collection, Observation & Culture

Plankton Collection

Collecting plankton samples can be as simple or complex as your time and resources allow. It involves passing quantities of sea water over or through a material of fine mesh to strain out planktonic organisms. The finer the mesh, the greater quantity of species you will collect. Most plankton samples are collected with a plankton net. Nets may be purchased commercially or homemade inexpensively. One type of homemade net can be prepared using a nylon stocking supported by a wire ring at the open end. A vial or jar should be inserted into a slit in the closed, or cod-end of the stocking. Make sure the mouth of the jar opens inside the stocking and the closed end of the jar hangs outside. This terminal container is where the plankton sample is concentrated for collection. The **cod-end bucket** (a term applicable to any terminal container) should be secured to allow removal with ease (e.g., using a rubber-band or hose clamp). If using a stronger net material, such as Nitex (used for most commercial nets), cod-end buckets can be made with heavier PVC. A threaded PVC link is permanently secured to the cod-end of the net. This permanent link allows the bucket to be easily attached and removed. The bucket is made from a short length of PVC pipe and threaded on one end; the threaded end joins with the threaded PVC link. The unthreaded end of the PVC pipe-bucket should be closed by gluing a flat piece of PVC or acrylic over the opening.

Commercial nets are available from many supply houses and vary in size from an open-end diameter of a few centimeters to over a meter. The length usually varies from less than a meter to several meters (for use in sampling deeper water). Cod-end buckets are usually fashioned from PVC and are proportional in volume to the net to which they are attached. The smallest nets may have a cod-end bucket with a volume of only milliliters, while large nets may have buckets with a capacity of several liters. Size and design of nets and their cod-end buckets are best specified by their intended use. For the novice, who is unsure of which size to use, a mouth diameter of 10 to 50 centimeters should collect plenty of plankton and not be too large or ungainly. Keep in mind that the diameter of the net and/or the time length of the haul determines the quantity of organisms you collect, not the size of the organisms. As has already been discussed, the size of the organisms collected is determined by the mesh size of the net.

The larger nets, of course, are intended to be towed behind a vessel in open water for a period of time which depends upon sampling goals. Smaller nets may also be towed horizontally (an *oblique* tow), or they may be hauled vertically between the water's surface and the ocean or bay floor. The latter is a convenient method for sampling from a wharf or pier and may also be used on a boat. The time length of the tow should reflect sampling goals and plankton population densities. Try a few minutes and observe the sample concentration. Slow, gentle tows will generally get more of a variety of organisms and keep more intact (e.g., for a horizontal dock-side sample: a casual walking pace or slower).

Collecting may also be done by means of a bottle or by use of a pump. In the former, a bottle may be lowered to depth on a line and the sample collected as a messenger, sent down along the cable, activates the bottle's shutting mechanism (professionally manufactured versions of this sampling apparatus are available). If a sample from the surface is desirable, one may simply fill any bottle from surface waters (recommended for an introduction to this type of sampling). Because of the small size of the sample, bottles are seldom used unless work is being done with protistans or bacteria. These organisms are present in high abundances that can be appropriately sampled with lower volumes. An electric pump can be used to collect samples from a known depth and with considerable ease and speed. A tube or pipe is lowered to a given depth and quantities of water pumped through a fine mesh net. Use of a rotary pump, rather than a piston type, is best. Pumps can be convenient, but note that organisms may be damaged by hydropressure or the impeller and organisms larger than the intake pipe diameter will not be sampled.

In some instances, it is desirable to disturb the specimen as little as possible. Diving, snorkeling or peering into the water wearing a face mask and collecting an undisturbed organism while it is still submerged is one way to accomplish this (Hamner, 1974). In this circumstance, the container with the organism can then be lifted from the water to the dock or boat.

Different net constructions have their own advantages and disadvantages. Two common options for net shape are the cone-shaped Standard net (Figure 1) or the modified Hensen net (Figure 2). The modification is the addition of the upper partial cone that surrounds the net opening. It is constructed of nonfiltering material, such as canvas or plastic, and is designed so that much of the water processed as the apparatus is drawn through the sea passes through the net, rather than regurgitating back out the net's mouth. The Standard net, without the Hensen modification, is said to be only one-tenth as efficient. The loss of samples by regurgitation can be lessened by drawing the net through the water at a reduced speed. As already mentioned, the slower the net can be towed or drawn, the greater number of individuals that will be retained within the net. A further modified net, the Helgoland type, is said to be another four times as efficient as the Hensen net.

We have presented here a few commonly used net designs, however, nets can be any shape, arranged in any configuration, and towed or drawn in any manner, as long as the sampling purpose of the investigator or student is accomplished.

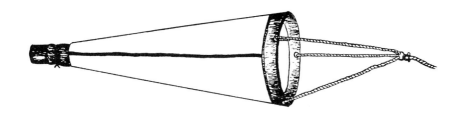

Figure 1. Standard plankton net.

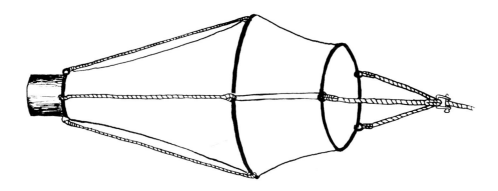

Figure 2. Hensen, modified plankton net.

Cod-end buckets may be flow-through (with portholes covered with plankton net material) or blind-ended. The blind cod-end, an example of which is described by Reeve (1981), is a cod-end bucket or bag that has no portholes or other outlets. Organisms concentrated in this container are less likely to suffer damage than if mesh-covered portholes are present. Delicate organisms can be destroyed or broken as they are forced against the mesh by filtering water.

Many of the line drawings in this guide were made of plankton collected with one of two nets from InterOcean of San Diego, California and are made by the Tsumuri-Seiki Kosakusho Co., Yokohama, Japan. Their Standard net, Marutoku B, has a mesh size of 106 µm and a net diameter of 45 cm. The Hensen type, Kitahara, has a mesh size of 8.2 µm and was used primarily for phytoplankton and protiste sampling. The latter net had a net diameter of 45 cm, as with the Standard net, but with an actual mouth of 24 cm due to the Hensen modification.

Various techniques may be used to sample at a given depth with a vertical closing net or other mechanical or automatic devices. There are also several types of continuous plankton recording devices. First developed by Hardy (1967), these are used for experiments where a large number of samples over a period of time is desirable. Using this method, samples are collected and preserved in a continuous, automatic operation.

A **flow meter** can be attached to many of the commercial nets to aid in calculating the volume of seawater processed by the net. This allows the researcher to estimate density of planktonic populations. A simple calculation can be made with any net used in a tow to give the volume of the cylinder of water that has passed through the net. Knowing this volume and the quantity of a species present in the sample, an estimation of the density of the population can be made. In calculating the volume of the cylinder of water processed, the radius of the mouth of the plankton net is used as the radius of the cylinder and the distance the net is hauled is the cylinder's height. This calculation assumes that all water is passing through the net and not being regurgitated, an assumption that is unreasonable if the net is towed too quickly or forcefully. For this reason, a calibrated flow meter is recommended for the most accurate volume determinations. However volume is calculated, determining field population densities is a worthwhile classroom exercise.

The following table (Table 1) is adapted from Wimpenny (1966). Understanding the relationship of mesh size (given here in microns because most commercial net mesh sizes are given in microns) and the silk number (for those making homemade silk plankton nets), may be helpful for conceptualization and silk selection. Note that these relationships may vary depending on the source used.

Table 1. Silk numbers and the corresponding inner distance of the mesh opening (μm).

Silk No.	Inside of mesh (μm)	Silk No.	Inside of mesh (μm)	Silk No.	Inside of mesh (μm)
0000	1,364	6	239	15	94
000	1,024	7	224	16	86
00	752	8	203	17	81
0	569	9	168	18	79
1	417	10	158	19	77
2	366	11	145	20	76
3	333	12	119	21	69
4	318	13	112	25	54
5	282	14	99		

A similar table from Newell & Newell (1967) gives the average diameter of the mesh aperture and varies slightly from the table above. Many nets are made of nylon which will withstand temperature and climatic severity as well as hard wear. They have the added advantages of more precise mesh and extremely small mesh openings.

Sample content will vary greatly, both in quantity and identity, depending upon the time of day (or night) and season of collection. This is not only due to differences in light and darkness, which might influence the vertical migration of some species, but also to tidal height and timing, species' blooms and population fluctuations, and to patchy spatial distribution. For a diversity of planktonic forms, Professor Smith has found samples to be best during daylight at high tide, especially when high tide is in the morning. Successive samples from the same location on a given day can result in increased species diversity.

There is considerable seasonal variation in the occurrence and abundance of plankton species as phytoplankton, holoplankton and meroplanktonic larval forms appear and disappear from the plankton. If the reproductive and spawning periods of benthic adult conspecifics of planktonic larvae are known, then the appearance of the larvae in the plankton may have a degree of predictability. Many marine invertebrates spawn in the Spring to early Summer. However, some species spawn in other seasons and may even spawn year round (M. Strathmann, 1987).

For classes interested in seeing diversity of taxonomy and form, weekly or monthly samples may show changes in species and forms as the composition of the plankton assemblage changes. Embryos and larvae in the plankton exhibit many different stages and forms, often within the same species.

Some specimens cling to the mesh (due to behavior or chance) as the water passes through the net. It is helpful to wash the net down from the outside. Similarly, when the apparatus is drawn from the water, it can be lowered several times to the rim of the net and washed up and down to dislodge clinging or hung-up organisms. This allows plankton to be washed down to the cod-end bucket below for collection. Organisms become concentrated in the cod-end. Professor Smith estimated the concentration factor in a 6 m haul with an average sized net and cod-end bucket to be about 20,000:1.

If the sample can be diluted during transport or while holding for later use, specimens will arrive in better condition. When too many organisms are crowded together, they use up available oxygen and suffocate. In the plankton tows from which many drawings in this guide were made, samples were taken directly to the laboratory where live animals were observed and illustrated. In this instance, dilution may be unnecessary. The live organisms can be transported to the classroom, provided the temperature is kept close to ambient. Seawater can be scooped up in a bucket where closed sample jars are then immersed in the water bath. This will maintain the samples within a few degrees of natural temperature for an hour or so, except during the hottest times of the year. Ice packed around jars can also be used to keep the sample cold until it arrives at the classroom or laboratory. Other problems resulting from crowding, other than respiration, should also be considered: crowding may affect predator-prey relationships and pollution by fecal pellets and excretion may reach lethal levels in a closed and overcrowded situation. Prolonged transportation will contribute to loss of the planktonic organisms in the sample.

The collection of live plankton is most desirable for classroom purposes. However, if preserving samples is necessary, the investigator may consult Steedman (1976) or Sournia (1978) for techniques on zooplankton and phytoplankton fixation and preservation techniques.

Plankton Observation

Observing living plankton is an exciting new adventure for most students and experienced biologists alike. Some techniques are best determined by trial and error. The variation in microscopy instruments and lighting allow for much experimentation. Transmitted light, by a sub-stage lamp and a frosted mirror, is good for observing the inner-workings of transparent species. If available, dark field, Nomarsky and phase-contrast microscopy methods can provide stunning views of living and preserved plankton (see Grimstone & Skaer, 1972, for a short discussion of these lighting techniques). Reflected light, from above the organism, will emphasize pigments and surface details when they are present. Experiment with and adjust lights, mirrors, condensers, etc., for maximum advantage. The following procedures may be helpful in the observation of plankton samples.

If possible, instructions for culturing plankton (below) should be followed for maintaining samples even if they are only needed for an hour or two in the classroom. Once bowls or jars containing the samples are in place in the laboratory or classroom, pipettes may be used to subsample from the bowl or jar. The student can then return to their microscope or desk with the subsample on a slide, in a watch glass, or in a finger bowl. Which of these containers to use and which type of microscope should be used (dissection or compound) depends upon the size of the organism and the visual detail desired.

After plankton has settled in bowls or jars, actively swimming organisms can be slurped from near the water's surface using a pipette (the size of the pipette determined from size of the organisms to be observed). Near the water's surface will be found most of the copepods, swimming veligers, medusae, ctenophores, siphonophores and a host of other free-swimming forms. Other organisms will be found in the settled plankton at the bottom of the bowl or jar. If sampling quantitatively, remember to check for organisms clinging or adhering to the walls of the container.

Many of the organisms will quickly become familiar due to their characteristic shape or motion. However, there is much to be learned about the biology of each organism in the sample. If one group or species is especially interesting, the investigator may seek out references included in this text and learn more about the identity and biology of these creatures. As samples are observed carefully, previously unnoticed and rare organisms will be discovered. It truly seems as if there will always be another undiscovered species in the sample.

For observation of bulky jellyplankton, the investigator may need depression slides, watch glasses or bowls. When viewing phytoplankton and protistans, a coverslip is recommended for more effective high power magnification.

For detailed observations of plankton specimens, a compound objective scanning lens of 4x or 5x is recommended. On a compound microscope, a 4x objective lens magnifies about 5 actual millimeters to fill your field of view (when using a 10x eyepiece). For a closer look, move to the 10x objective lens. Care should be taken to avoid getting seawater on the lens. In the event this happens, swab the lens generously with manufacturer-recommended lens cleaner and dry with lens paper.

Plankton Culture

If plankton is to be kept alive long-term once in the laboratory (for more than just an hour or two) the sample should be kept diluted and cool. If a sea table with flowing seawater is available, this is ideal. Bowls or jars with the diluted samples should be half-submerged in the cool seawater. Otherwise, keeping samples on ice can keep organisms alive. The goal is to maintain the temperature at approximately that which the organisms are accustomed to. Where an aquarium is available, the sample can be immersed or suspended in the tank, preferably with the top off to allow for diffusion of oxygen. For example, the sample can be half-submerged in a shallow frog tank in which tap water is constantly running. This may keep the sample cool enough to maintain it a day or two. An aquarium with a thermostat which can be set to the proper temperature will work well. The samples are immersed in the aquarium, supported to keep the top above the water line, with lids removed. Following this procedure, samples have been kept for several days. However the problem of live plankton maintenance is approached, many plankton die quickly when removed from their natural environment. A good general rule is to observe and identify the organisms as quickly as possible after the sample has been collected.

When culturing organisms for weeks or longer, crowding and temperature are still the critical factors. In some cases, it is desirable to keep the organisms suspended. Prize specimens can be isolated in separate containers with plenty of seawater. The water-bath method, analogous to that described for maintaining the entire plankton sample for classroom observation, works well. Other methods of separating and maintaining selected plankton specimens include the stopped vial and ice cube tray methods. Selected specimen or specimens placed in a test tube or a small vial can be stoppered and allowed to float in an aquarium. Isolated specimens can be placed in the individual compartments of an ice cube tray, or a similar compartmented plastic box, and allowed to float on the water's surface. With any method, the more organisms and less seawater volume you have, the more often the water should be changed.

Filtered seawater lessens the risk of contamination by unwanted microorganisms. An artificial seawater may also be used. Several seawater formulae are published; Cavanaugh (1975) and Bidwell & Spotte (1985) explain a general purpose seawater formula developed at the Marine Biological Laboratory of Woods Hole, Massachusetts for use in marine aquaculture.

All samples should be kept at normal sea temperature. A water bath for the jar, beaker or bowl is good for this. Some researchers prefer to keep plankton in small watch glasses. This allows organisms to be found again easily. In the case of meroplanktonic larvae, some may settle and attach to the substratum. If they settle and metamorphose in a watch glass, it can be placed under a dissection microscope and the settled individual observed.

If cultures are to be maintained for more than a few days, then feeding them may be necessary. Food cultures can be prepared from a variety of algae, diatoms, protozoa or small invertebrate larval forms. Cultures of the phytoplankton genera *Dunaliella, Isochrysis, Monochrysis, Phaeodactylum, Sceletonema, Chaetoceros, and Rhodomonas* are some examples of commonly used zooplankton food. For suggestions on obtaining and maintaining algal cultures see Andersen et al., 1991.

A swinging paddle system, described by R. Strathmann (1971) and detailed by M. Strathmann (1987), has become a common method of keeping plankton suspended in culture. If suspension is desired, any method which stirs the water without damaging the organism is satisfactory.

Culture methods for many of the invertebrates found in plankton samples are detailed in M. Strathmann (1987).

Chapter 2
Quick Flip Reference

This relatively small volume covers most of the Kingdoms and Phyla represented in the plankton. It does not, however, provide a comprehensive guide to identifying species. Some species may be identified with this guide because they are either morphologically unique, or are regionally the only representative of the taxonomic group. In most cases, organisms illustrated in this book are identified to family or genus. Since many genera and families are world-wide in distribution, the necessary nonspecific identifications in this volume keep this guide applicable to many global regions.

Part of the intrigue and appeal of marine plankton is the diversity at all taxonomic levels. Forms appearing very similar may be different species. Some species are cryptic, being indistinguishable by visual examination. In many taxonomic groups (e.g., copepods, polychaetes, diatoms, etc.) there are many species of similar morphology. Identification of these species becomes complicated and, in some cases, may not be possible with currently available descriptions. Even so, for a single group (e.g., copepods) there exist volumes of descriptive literature easily exceeding the thickness of this general text. Marine biologists, specializing in the study of a particular group, are familiar with the detailed taxonomic keys required for that group. However, few are extensively familiar with the broad spectrum of plankton species.

It is the intent of this manual to provide simple identifications of plankton groups based upon easily recognizable general patterns of morphology. It has been the experience of the author and others that, for general identification in a book as small as this volume, a dichotomous key can be more confusing than helpful. The following key is very simply a quick visual guide to which the investigator can refer in order to learn the general classification of their organism. Once the general classification is known, the student/researcher turns to the section of the guide describing various members of the taxa in more detail. The location of the pertinent

Quick Flip Reference

sections are indicated by the plate numbers underneath the illustrations in the quick flip reference below. Illustrations below will not identify all unknown organisms being investigated, but should give a good indication of where to look in the main body of this text to identify organisms similar in attributes to those illustrated below. If your mystery plankton is morphologically unusual for its group, perhaps telltale characteristics, such as setae or eyespots, will lead you to the correct section. If all fails, flip through the book. Chances are you will quickly find something similar to your organism.

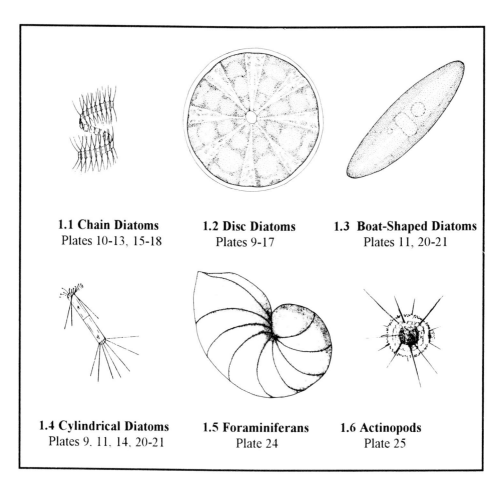

1.1 Chain Diatoms
Plates 10-13, 15-18

1.2 Disc Diatoms
Plates 9-17

1.3 Boat-Shaped Diatoms
Plates 11, 20-21

1.4 Cylindrical Diatoms
Plates 9, 11, 14, 20-21

1.5 Foraminiferans
Plate 24

1.6 Actinopods
Plate 25

Plate 1. Quick Flip Reference: Diatoms, Foraminiferans & Actinopods

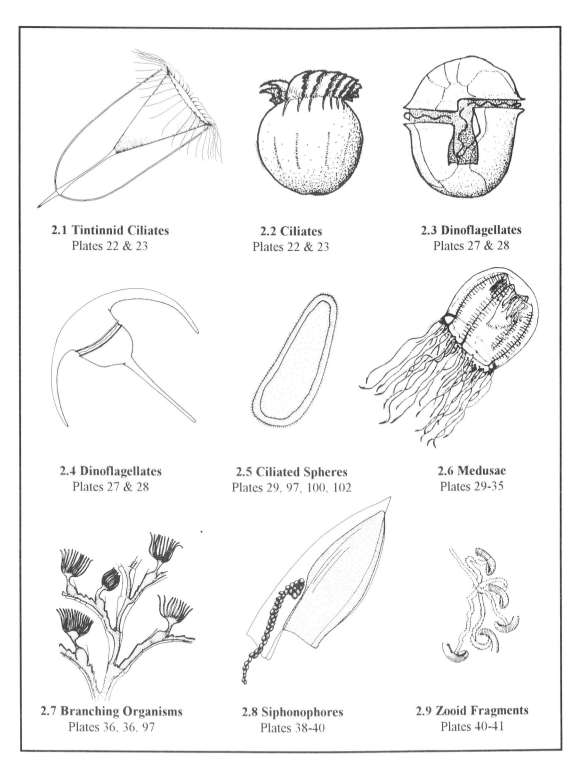

Plate 2. Quick Flip Reference: Ciliates, Dinoflagellates, Cnidarians & Miscellaneous

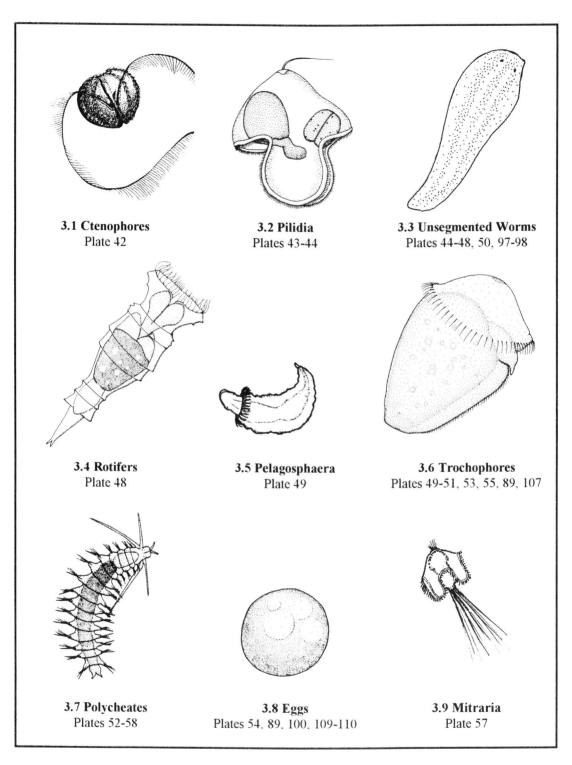

Plate 3. Quick Flip Reference: Ctenophores, Pilidia, Vermiformes, Eggs

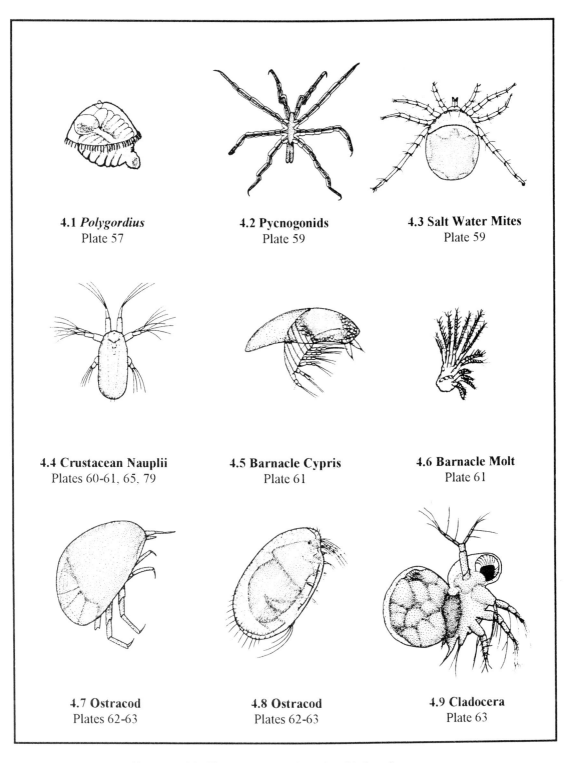

Plate 4. Quick Flip Reference: *Polygordius*, Various Crustaceans

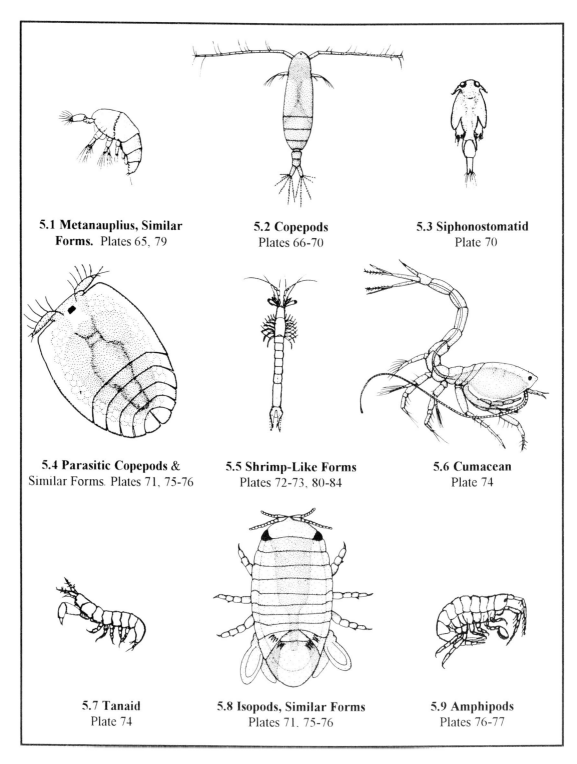

Plate 5. Quick Flip Reference: Various Crustaceans

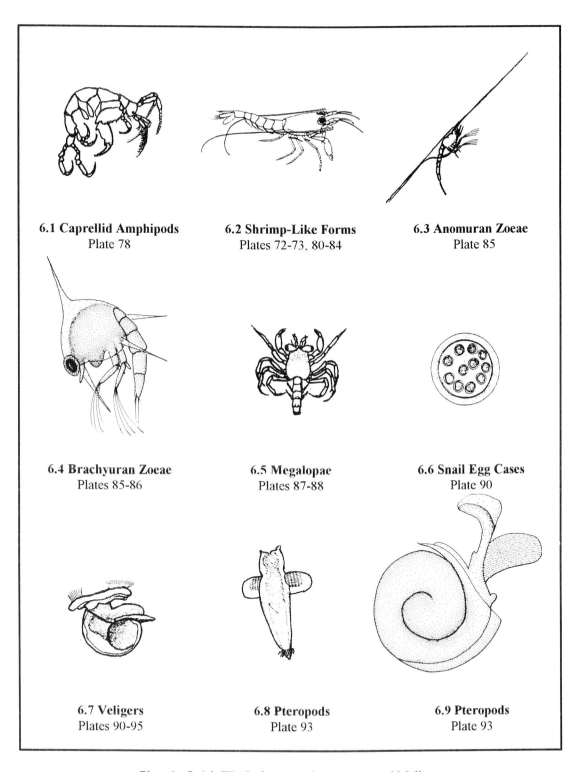

Plate 6. Quick Flip Reference: Crustaceans and Molluscs

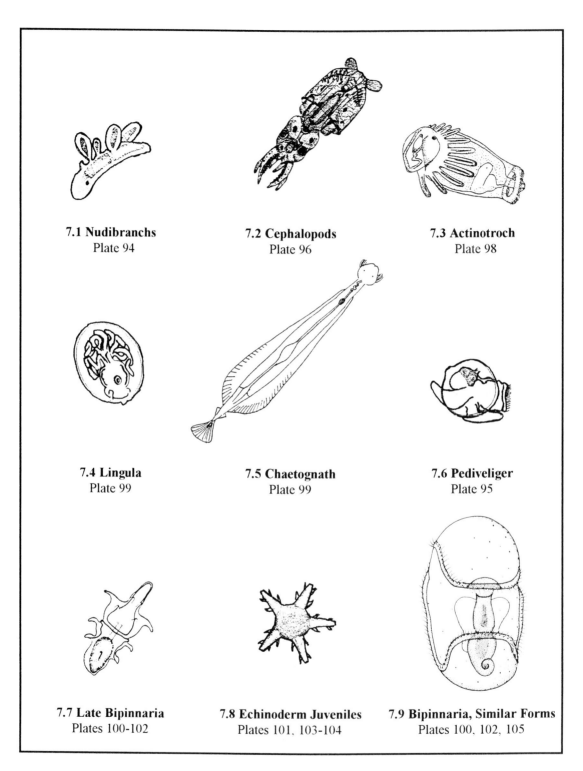

Plate 7. Quick Flip Reference: Molluscs and Deuterostomes

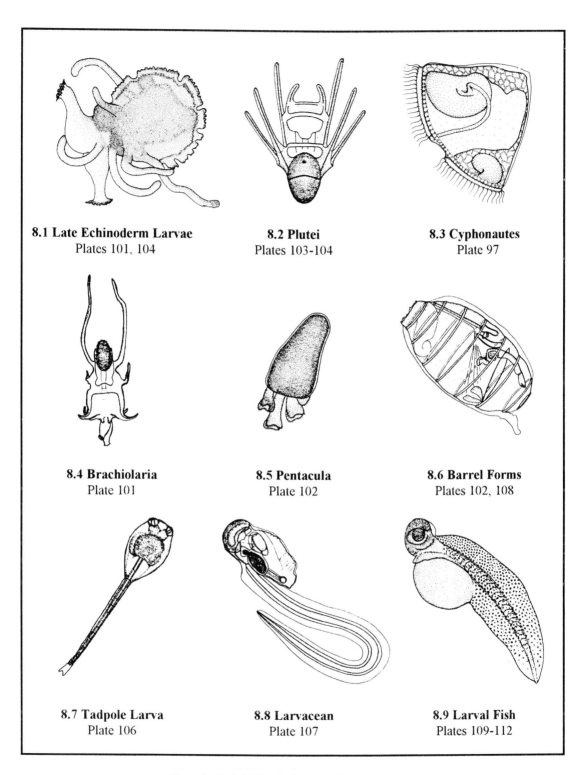

Plate 8. Quick Flip Reference: Deuterostomes

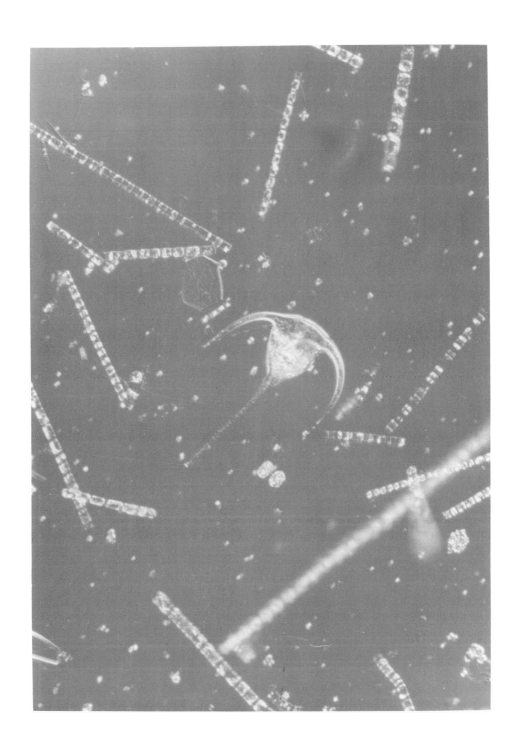

Chapter 3
Planktonic Protistans

Finding two references which completely agree upon the proper system of taxonomic classification for the Protista is difficult. The tendency to assume that morphologically similar groups are related has led to the grouping of taxa when they may not be related at all. With the advent of **molecular systematics** and other techniques for determining evolutionary relationships, it is becoming increasingly clear that true **phylogenetic relationships** among the **eukaryotic** microorganisms are uncertain. For now, organisms of known affinity are placed together in a phylum, while the collection of these phyla are lumped into the Kingdom **Protista** (also called Protoctista). Margulis et al. (1990) define the Protista as "the grouping of eukaryotic microorganisms and their descendants exclusive of animals, plants and fungi." Therefore, the protistans include the algae (a generic term including flagellates), ciliates, foraminifera, various molds and some other microorganisms. Different authors and researchers use different systems when naming and defining the phyla. We use the Protista classification system defined by Margulis et al. (1990).

Many protistan phyla are represented in planktonic marine communities. Planktonic protistans are a very diverse group of organisms whose presence and production are of utmost importance to the ocean's ecology. Some larger protistans, many of which are mobile, can be captured when towing for larger zooplankton. Because of their ecological importance and the need to be able to identify them in plankton tows, this chapter provides general identifications of common planktonic protistans. However, this guide is geared primarily towards the identification of marine coastal zooplankton. This chapter is included to help the investigator of zooplankton (meaning plankton from the Kingdom Animalia) determine if organisms being observed are protistans or unfamiliar animals. Descriptions in this chapter, intended to give brief representation of common marine phyla, will help the investigator approximate the identity of planktonic protistans. Additional references to aid in identification are sometimes given with the information on specific groups below. Margulis et al. (1990) provides identification of protistans in general and is an excellent source for references to more detailed and technical keys. Lee et al. (1985) gives an illustrated guide to those protistans with animal-like characteristics.

Phytoplankton - Pastures of the Ocean

The term phytoplankton is a generic term for planktonic organisms which **photosynthesize**. Which organisms are classified as phytoplankton often depends upon the taxonomist. Definitions become especially difficult when one considers the **flagellates**. Some flagellates photosynthesize while others are primary or secondary **consumers**. In any case, the larger phytoplankton caught in plankton nets consist primarily of the **diatoms** and dinoflagellates, though other flagellated species and fragments of macroscopic red, green, and brown algae are often found. Diatoms, and many dinoflagellates, are known for being **autotrophic**, or having the ability to produce their own food through photosynthesis. Autotrophs require only sunlight and dissolved nutrients, including CO_2 from the surrounding seawater, in order to grow. Other important nutrients for primary production include phosphates, nitrates, silicates, and varied proportions of minor elements.

The key ecological role of the phytoplankton is the primary production of **biomass** and oxygen through the process of photosynthesis. Many think that phytoplankton play a key role in the removal of CO_2 from the earth's atmosphere. In fact, the globe's polar oceans, where primary productivity is high, may be a major sink for CO_2. Photosynthesis converts the raw material of CO_2 and H_2O into simple sugars with a photochemical reaction. These sugars can be converted to complex substances and incorporated into the phytoplankton's cell body or stored for future use. The resultant growth and reproduction of phytoplankton cells is primary production.

Energy from sunlight, required for photosynthesis, is available to primary producers only within the upper 100-200 meters of the ocean. This euphotic zone may be as shallow as several meters in cases of extreme turbidity. The depth of the euphotic zone is dependent upon the incidental angle of the sunlight, the clarity of the atmosphere, and the turbidity of the seawater.

Some of the ocean's CO_2 is derived from the atmosphere (both natural and anthropogenic origins). In addition to these sources, dissolved CO_2 comes from respiration of marine life and decomposition of organic matter in the ocean. As organic matter decays, it often sinks into the ocean's depths. Essential nutrients, released into the ocean by the breakdown of organic matter, are therefore often found in deeper waters nearer to the ocean floor. Primary producers usually require these nutrients to be present in the upper layers of the ocean, where light is available, in order to photosynthesize. Uniting the euphotic zone with nutrient-rich waters requires vertical mixing of the ocean. A common and important way that mixing takes place is via **upwelling**. Upwelling is the movement of deep, cold, nutrient-rich waters towards the ocean's surface and often occurs around coastlines. This is due to the headlands, subtidal topography, storms, and prevailing offshore currents that are common to these regions. Areas of persistent upwelling support floral and faunal communities of great diversity and biomass. For example, the abundant sea life of Monterey Bay, California, depends much upon natural upwelling from the bay's submarine canyon. In Monterey Bay, as with many coastal environments the world around, as cold, nutrient-rich water is upwelled, local phytoplankton populations may explode into what is referred to as a **bloom**. These blooms, occurring when conditions are optimal for phytoplankton growth, provide the primary production that fuels entire marine ecosystems.

Copepods, perhaps the most abundant animals in the sea, are probably the largest consumers of diatoms and other phytoplankton. Many meroplanktonic larvae and other planktonic invertebrates also feed upon primary producers such as diatoms. When the energy from the egg yolk of developing invertebrates is depleted, many **planktotrophic** larvae turn to a strategy of planktonic feeding.

Much of the research on phytoplankton is on their taxonomy and biological diversity, the role of smaller plankton in the microbial food web (Sherr & Sherr, 1991), cycles of primary production measured with the use of satellite imagery, the predatory behavior of heterotrophic dinoflagellates, and the production of toxins which enter the food web. Much research is also being done concerning phytoplankton's role in removing CO_2 from the atmosphere. Some scientists, studying a phenomenon termed **global warming**, feel that identifying sources and sinks of CO_2 is the key to understanding the trend of warmer atmospheric temperatures in the latter part of this century. Global warming is a general atmospheric trend towards warmer temperatures, which some feel is geologically recent and possibly due to the by-products of human civilization. Because of the aforementioned potential for removing CO_2 from the atmosphere, much important research is being conducted on phytoplankton. Research efforts are also intense in the area of mariculture. For example, phytoplankton can be used to feed and raise fishes and shellfishes at underwater farms. There are many other areas of scientific study where phytoplankton are the focal point. It is certain that we are just beginning to investigate the importance and potential of primary production in the ocean.

Planktonic Protistans

The protistan groups commonly observed in plankton samples include the diatoms (phylum Bacillariophyta), the **ciliates** (phylum Ciliophora), the **foraminiferans** (phylum Granuloreticulosa), the **radiolarians** (phylum Actinopoda) and several **flagellate** phyla. Among the many flagellate taxa that are found in marine plankton samples are the following phyla: Cryptophyta, Prymnesiophyta, Chrysophyta, and Chlorophyta. Dinoflagellates (phylum Dinoflagellata or Dinomastigota) are another flagellate group that are common and important in planktonic marine ecosystems.

Because protistans are numerous and diverse, phyla and classes are usually treated singularly in identification publications. The following is not intended to be a comprehensive overview, rather it should familiarize the investigator with general forms and give them a starting point for further research on protistans. The phyla described below are not all of the phyla represented in marine plankton, but are those that will be regularly encountered in classroom situations. Please consult the references suggested in this chapter for more detailed information on identifying protistans.

Diatoms

The diatoms (a few representatives of which are illustrated in Plates 9-21) belong to the phylum Bacillariophyta and are often called the golden-brown algae. At high magnification (i.e., 450X), **chloroplasts**, where photosynthesis takes place, can be observed in the cells of diatoms. The diatom cell consists of an outer case or covering known as a frustule and an inner protoplast layer. The frustule is composed of silica and divided into two halves, or valves, which connect with one slightly inside of the other forming a capsule. The slightly larger valve with its bands is the **epitheca**, while the smaller valve is the **hypotheca**. The circumference delineated by the valve seam is called the girdle. The outer surface of the capsule is often etched with intricate geometric patterns. Some diatoms are single-celled, while others form cell colonies which are chain-like. Diatoms display a diversity of cell shapes. Some are dramatically flattened (e.g., *Coscinodiscus*, Figure 9.4), yet others are elongated (e.g., *Rhizosolenia*, Figure 11.4).

The inner protoplast adheres to the inner wall of the frustule with interconnecting strands which, at times, support the nucleus of the cell midway between the valves. Chloroplasts are usually evident, especially when the cell is illuminated with epifluorescence. Some species possess a definitive number of chloroplasts per cell. Cell division occurs mitotically. During this process, the two parent valves, both now to be epithecae, cradle the developing hypothecae. The protoplasts for the two new cells are slightly reduced in size when compared to the pre-division protoplast. In some cases, a sexual, or pseudo-sexual, form of reproduction will produce an **auxospore**. The auxospore is an expanded protoplast that forms a primary cell of maximum size. Strategies of sexual reproduction in diatoms are diverse.

Diatoms are often identified by the shape of the valve (face) of the diatom. There are three major shapes of diatom valves: disc-like, rod-like, or boat-shaped. The diatoms with disc-like valves exhibit a valve ornamentation that radiates from a point (e.g., *Coscinodiscus, Melosira,*

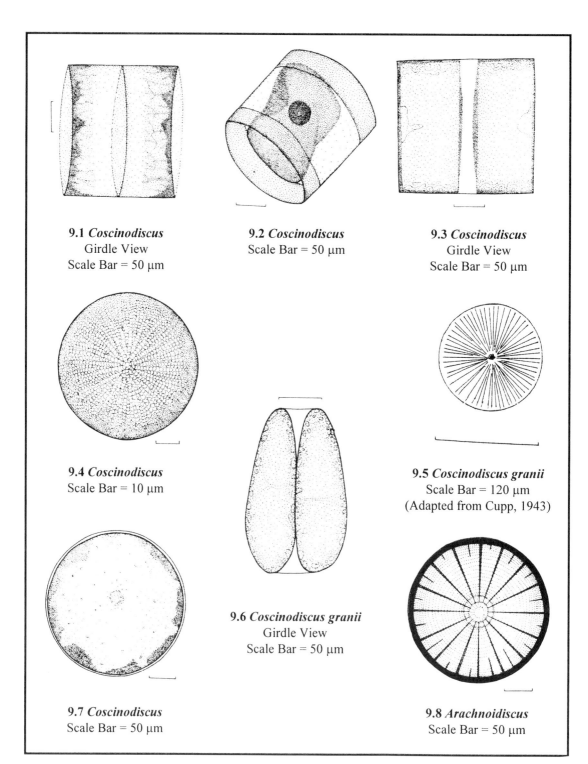

Plate 9. Diatoms, Class Coscinodiscophyceae

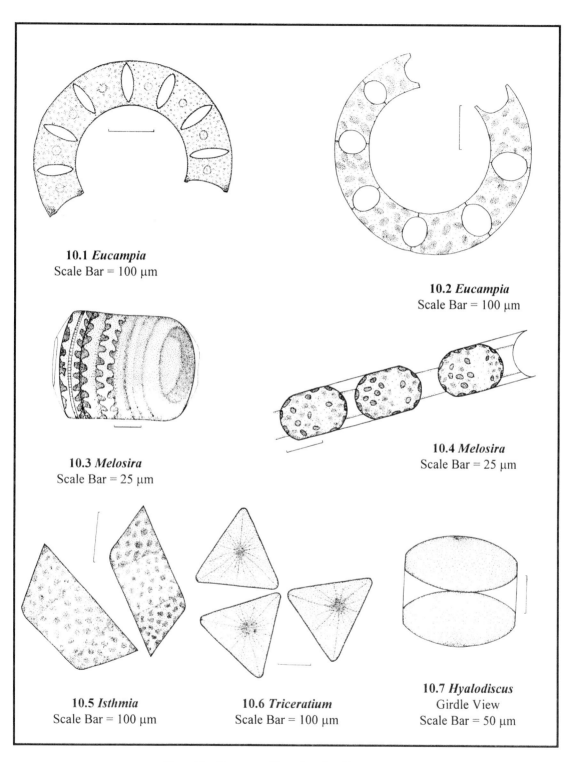

Plate 10. Diatoms, Class Coscinodiscophyceae

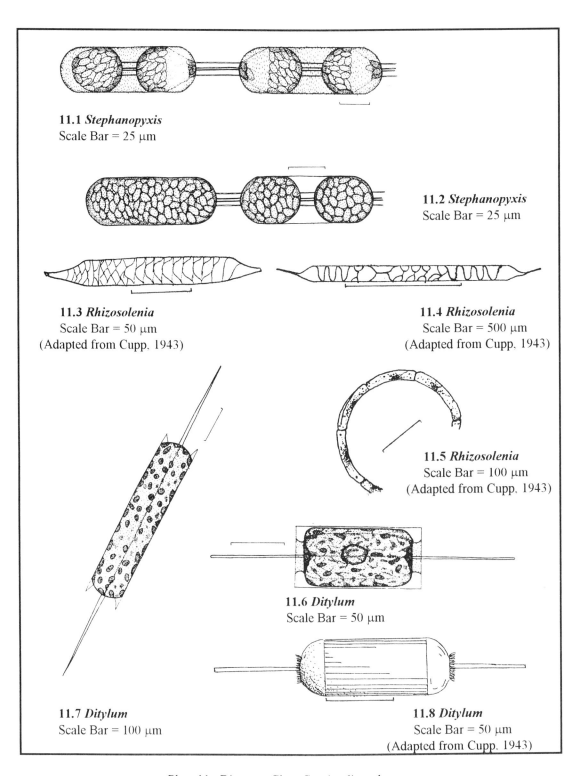

Plate 11. Diatoms. Class Coscinodiscophyceae

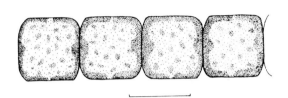

12.1 *Lauderia*
Scale Bar = 25 μm

12.2 *Biddulphia*
Scale Bar = 25 μm

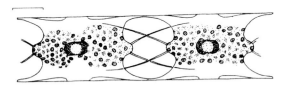

12.3 *Odontella*
Scale Bar = 25 μm

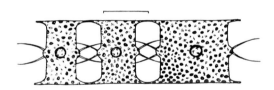

12.4 *Odontella*
Scale Bar = 50 μm

12.5 *Odontella*
Scale Bar = 50 μm

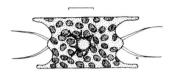

12.6 *Odontella*
Scale Bar = 25 μm

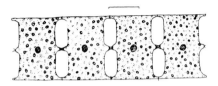

12.7 *Odontella*
Scale Bar = 50 μm

Plate 12. Diatoms, Class Coscinodiscophyceae

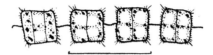

13.1 *Thalassiosira* Scale Bar = 50 μm (Adapted from Cupp, 1943)

13.2 *Thalassiosira* Scale Bar = 25 μm

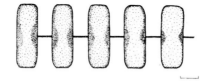

13.3 *Thalassiosira* Scale Bar = 25 μm

13.4 *Sceletonema* Scale Bar = 10 μm (Adapted from Cupp, 1943)

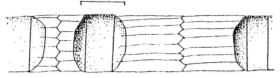

13.5 *Sceletonema* Scale Bar = 10 μm (Adapted from Cupp, 1943)

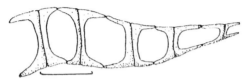

13.6 *Climacodium* Scale Bar = 100 μm (Adapted from Cupp, 1943)

Plate 13. Diatoms, Class Coscinodiscophyceae

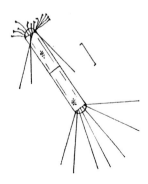

14.1 *Corethron*
Scale Bar = 50 µm
(Adapted from Cupp, 1943)

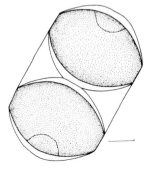

14.2 *Hyalodiscus*
Scale Bar = 50 µm

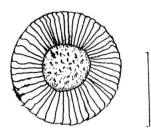

14.3 *Planktoniella*
Scale Bar = 100 µm
(Adapted from Vineyard, 1975)

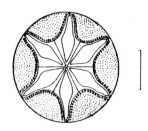

14.4 *Asterolampra*
Scale Bar = 10 µm
(Adapted from Cupp, 1943)

Plate 14. Diatoms, Class Coscinodiscophyceae

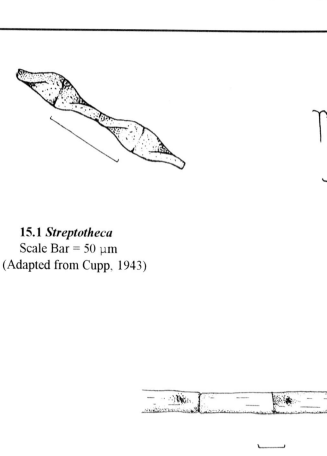

15.1 *Streptotheca*
Scale Bar = 50 µm
(Adapted from Cupp, 1943)

15.2 *Schröderella*
Scale Bar = 25 µm
(Adapted from Cupp, 1943)

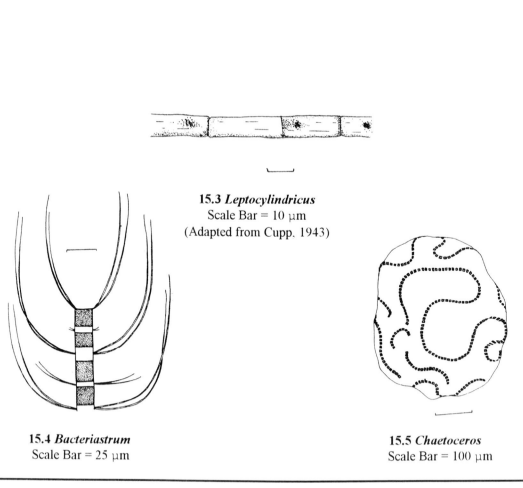

15.3 *Leptocylindricus*
Scale Bar = 10 µm
(Adapted from Cupp, 1943)

15.4 *Bacteriastrum*
Scale Bar = 25 µm

15.5 *Chaetoceros*
Scale Bar = 100 µm

Plate 15. Diatoms. Class Coscinodiscophyceae

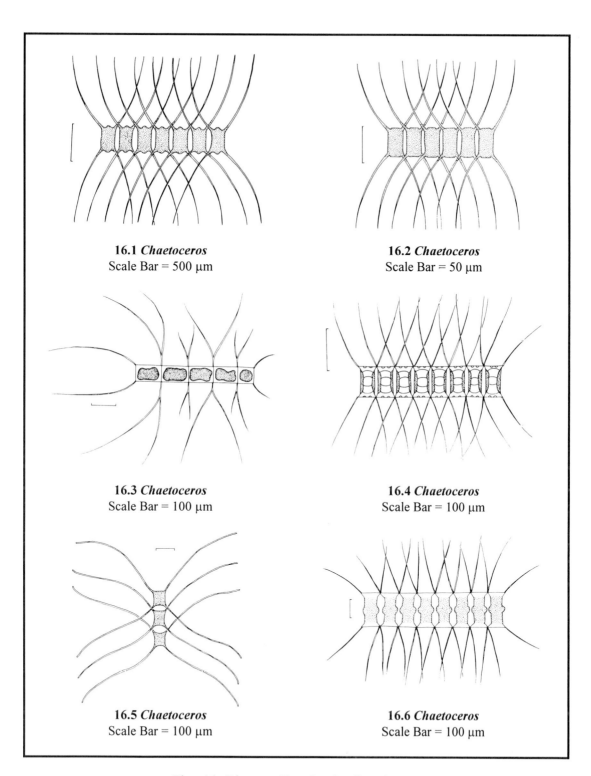

Plate 16. Diatoms, Class Coscinodiscophyceae

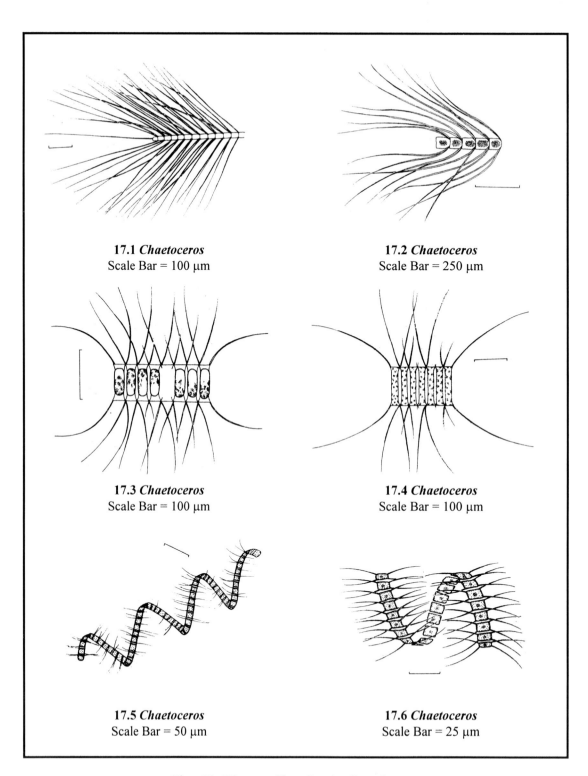

17.1 *Chaetoceros*
Scale Bar = 100 μm

17.2 *Chaetoceros*
Scale Bar = 250 μm

17.3 *Chaetoceros*
Scale Bar = 100 μm

17.4 *Chaetoceros*
Scale Bar = 100 μm

17.5 *Chaetoceros*
Scale Bar = 50 μm

17.6 *Chaetoceros*
Scale Bar = 25 μm

Plate 17. Diatoms, Class Coscinodiscophyceae

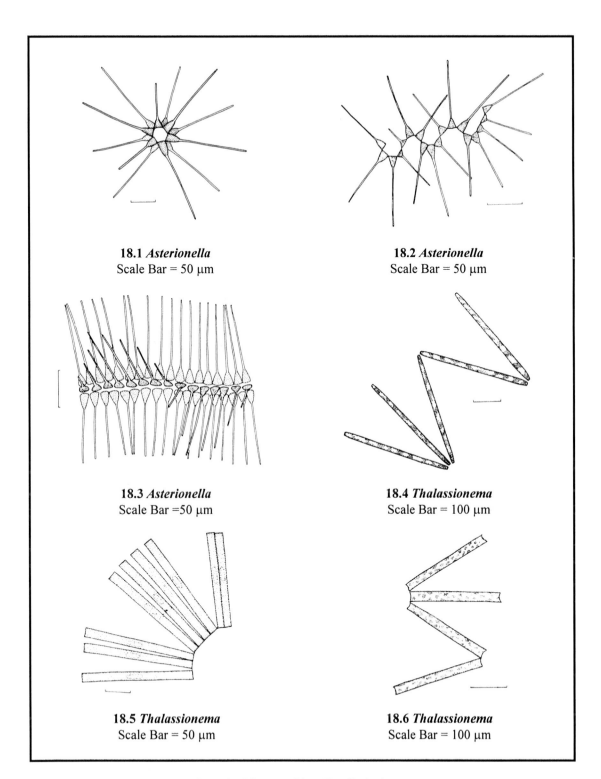

Plate 18. Diatoms, Class Fragilariophyceae

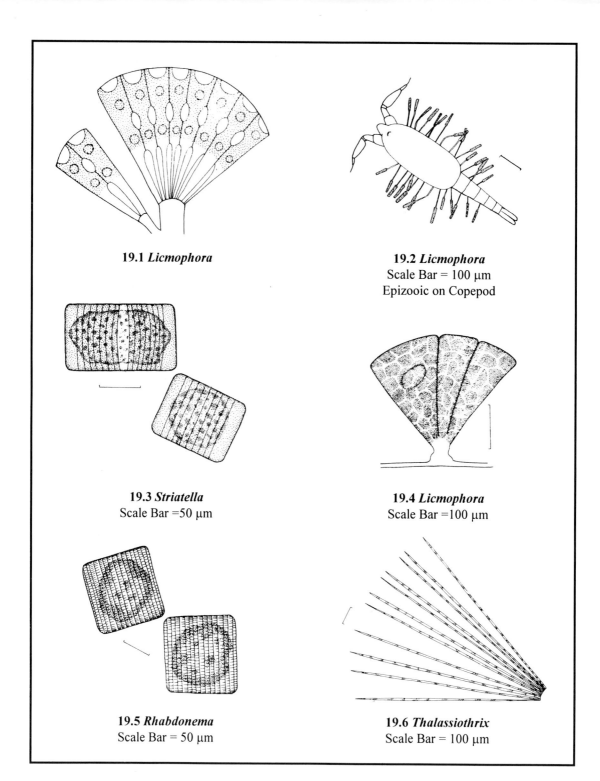

Plate 19. Diatoms, Class Fragilariophyceae

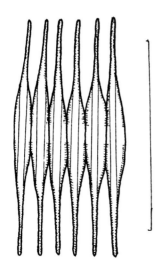

20.1 *Fragilaria*
Scale Bar = 50 µm
(Adapted from Cupp, 1943)

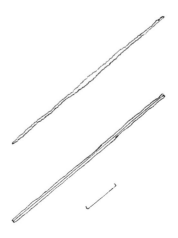

20.2 *Synedra*
Scale Bar = 100 µm
(Adapted from Cupp, 1943)

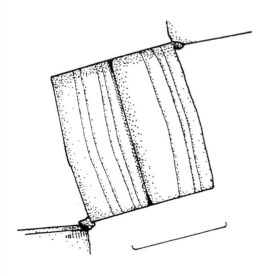

20.3 *Grammatophora*
Scale Bar = 100 µm
(Adapted from Round & Crawford, 1990)

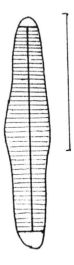

20.4 *Grammatophora*
Scale Bar = 100 µm
Valve View

Plate 20. Diatoms, Class Fragilariophyceae

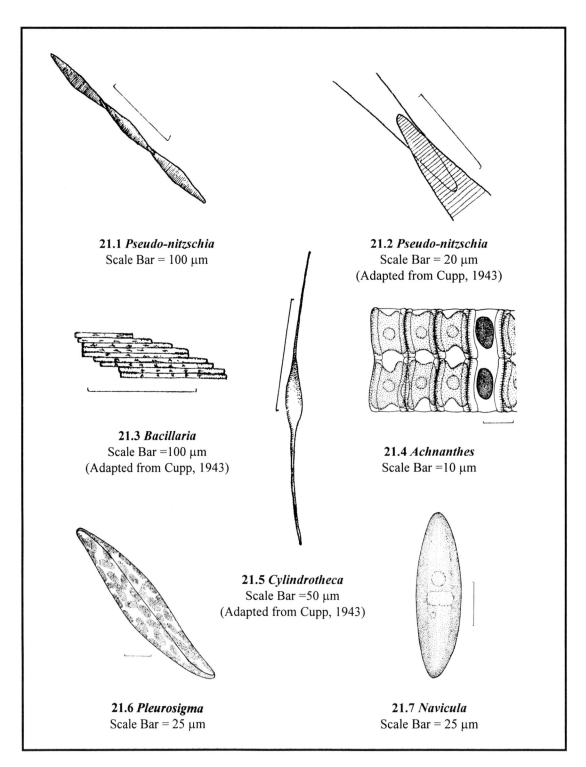

Plate 21. Diatoms, Class Bacillariophyceae

and *Chaetoceros*, all in the Class Coscinodiscophyceae, Plates 9-17). Colonies of disc-shaped diatoms may be held together by threads, gelatinous matrices, or siliceous extensions of the valves. The rod-like, linear valve shape is typical of diatoms from the Class Fragilariophyceae (e.g., *Thalassionema, Licmophora,* and *Rhabdonema,* Figures 18.5, 19.4 & 19.5, respectively). Colonies of rod-shaped diatoms are often held together by mucous pads or the faces of the valves themselves. The nonmotile Fragilariophyceae are pennate diatoms with many similarities to the centric diatoms. The Bacillariophyceae, many of which are boat-shaped (e.g., Plate 21) are also pennate diatoms. The Bacillariophyceae, however, possess a slit or raphe structure which allows movement. Colonies of these moving diatoms are often held together at the overlapping tips of neighboring cells. With each cell pushing against its neighbor, the entire colony appears to move in a coordinated fashion.

Though published in the first half of the twentieth century, some of the earliest guides to diatom identification remain important works. This includes Hustedt (1927-1966) and Cupp (1943). A more modern and less involved work on the diatoms of the West Coast of North America is Vinyard (1975).

Ciliates

The most common marine planktonic ciliates (phylum Ciliophora) found in net tows for larger zooplankton are those from the class Spirotrichea, especially oligotrichs and choreotrichs (subclass Choreotrichea). A small sample of some planktonic marine ciliates is illustrated in Plates 22 & 23. Usually the most notable are the tintinnids, a ciliate group possessing an outer skeleton or **lorica** (often shaped like a cone or a wine glass). When disturbed, some species with loricae retract or withdraw into the skeleton for protection.

As the name implies, all ciliates possess cilia, which is their mode of locomotion. Ciliary movement is a quick, pulsating, rhythmic motion, both forward and back. Using cilia to cruise through the water, tintinnids and other marine ciliates are very fast for their size. If a very small organism is zooming around on the slide too quickly to

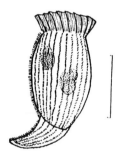

22.1 *Strobilidium*
Order Choreotrichida
Class Spirotrichea
Scale Bar = 10 µm
(Adapted from Montagnes & Taylor, 1994)

22.2 *Strombidinopsis*
Order Choreotrichida
Class Spirotrichea
Scale Bar = 50 µm
(Adapted from Montagnes & Taylor, 1994)

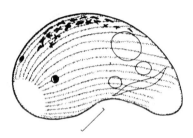

22.3 *Chlamydodon*
Class Phyllopharyngea
Scale Bar =25 µm

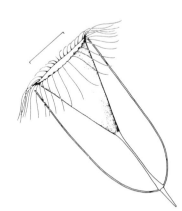

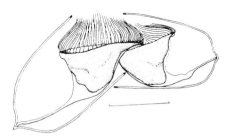

22.4 *Parafavella*
Class Spirotrichea
Scale Bar = 50 µm

22.5 *Parafavella*
Fission
Scale Bar = 100 µm

Plate 22. Ciliophora (Ciliates)

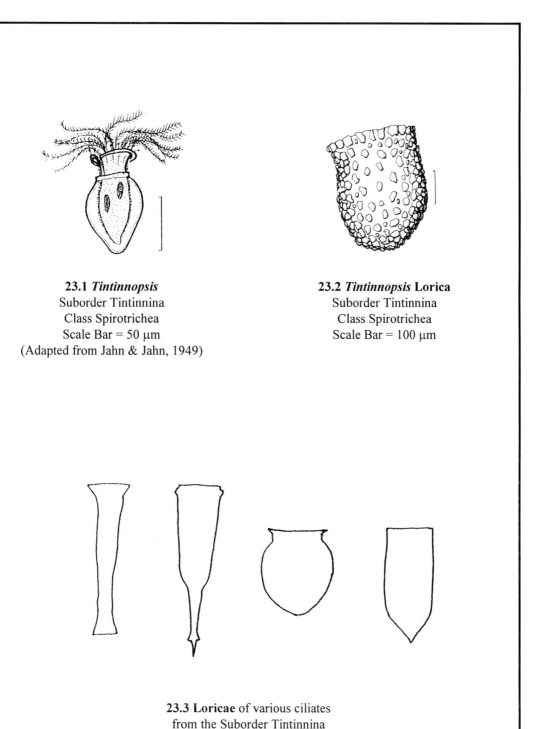

23.1 *Tintinnopsis*
Suborder Tintinnina
Class Spirotrichea
Scale Bar = 50 µm
(Adapted from Jahn & Jahn, 1949)

23.2 *Tintinnopsis* Lorica
Suborder Tintinnina
Class Spirotrichea
Scale Bar = 100 µm

23.3 Loricae of various ciliates
from the Suborder Tintinnina

Plate 23. Ciliophora (ciliates), tintinnids and loricae

identify, chances are it is a ciliate. Some stalked ciliates are benthic and sessile, though these still may be detached and swept into the water column where they can be captured in a plankton tow.

Ciliates are a diverse group. As we have discussed, the choreotrichs (including tintinnids) and oligotrichs are among the more common planktonic marine ciliates. Marshall (1969) gives a good overview of the types of tintinnids that may be encountered. Photographs and explanations of taxonomy in five oligotrich ciliates are provided in Montagnes & Taylor (1994).

Foraminiferans

Foraminiferans (phylum Granuloreticulosa, class Foraminifera) have a **test**, which is often calcareous and resembles a microscopic nautilus shell. Through assorted holes in the test, **cytoplasm** comes through to form a thin outer layer. From this layer arises an impressive network of **reticulopodia**. In planktonic foraminiferans these reticulopodia are probably used for suspension and capturing/engulfing food particles. Foraminiferans are generally known for the accumulation of their calcareous tests, after their deaths, in ocean sediments (called **foraminiferan ooze**). When huge populations of foraminiferans living in the water column die off, empty tests sink to the sea floor and pile up in a sediment layer. These fossil layers have aided geological studies of ocean sediments. A few common genera of planktonic marine foraminifera are illustrated in Plate 24. A helpful key to the identification of the Foraminifera is given in Lee et al. (1985).

Actinopods

Marine actinopods include the classes Acantharia, Phaeodaria and Polycystina. Polycystines and phaeodarians are commonly referred to as radiolarians. Acantharian skeletons are made of strontium sulfate, while

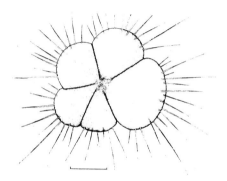

24.1 *Globigerina*
Family Globorotaliidae
Scale Bar = 25 µm

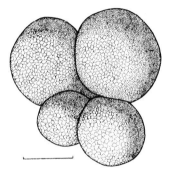

24.2 *Globigerina bulloides*
Family Globorotaliidae
Scale Bar = 25 µm

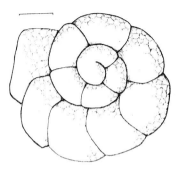

24.3 *Discorbis*
Family Discorbidae
Scale Bar = 25 µm

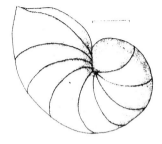

24.4 Unidentified Foraminiferan
Family Nonionidae
Scale Bar = 25 µm

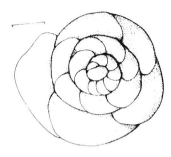

24.5 Unidentified Foraminiferan
Family Trochamminidae
Scale Bar = 25 µm

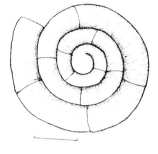

24.6 Unidentified Foraminiferan
Family Epistomariidae
Scale Bar = 25 µm

Plate 24. Granuloreticulosa (foraminiferans)

25.1 Radiolarian Test
(Adapted from Buchsbaum, 1976)

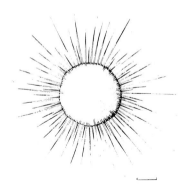

25.2 *Orbulina*
Radiolarian
Scale Bar = 25 μm

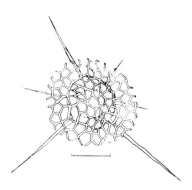

25.3 *Pleuraspis*
Acantharian
Scale Bar = 25 μm

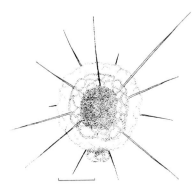

25.4 *Acanthometra*
Acantharian
Scale Bar = 25 μm

Plate 25. Superclass Actinopoda

radiolarian skeletons are siliceous in nature. Many actinopods have the general form of a ball with long, slender spines covered with cytoplasm and called **axopodia**. The spines radiate from the **central capsule**, which the cytoplasm surrounds in a foamy-looking mass called the **calymma**. Spines provide greater surface area for capture and ingestion of food by the axopodia, as well as better flotation. Similar to foraminiferan oozes, **radiolarian oozes** cover the sea floor in some areas. While foraminiferan tests tend to be calcareous and dissolve at great depths, the siliceous skeletons of radiolarians do not dissolve with pressure and can be found in the deepest parts of the ocean. A few common genera of planktonic marine Actinopoda are illustrated in Plate 25.

Flagellates

Flagellates are an immensely diverse group and might be a lifetime study in themselves. All flagellates have at least one **flagellum**. Some flagellates have chloroplasts and some do not. Of those flagellates which possess chloroplasts, most use the photosynthetic pigment chlorophyll *a*. Some autotrophic flagellates have other chlorophyll pigments, carotenes, and/or xanthophylls in addition to chlorophyll *a*. Some flagellates have the characteristics of animals, some have plant characteristics and others possess characteristics of both groups. Taylor et al. (1987) have published a discussion of the controversy and confusion in flagellate nomenclature. Plate 26 shows representative genera of four phyla of flagellates common in plankton samples. Some dinoflagellates, an ecologically important and morphologically varied phylum, are illustrated on Plates 27 & 28.

Dinoflagellates possess two flagella, found in grooves or trenches in the cell wall. Reproduction in dinoflagellates can be both sexual or asexual. Some toxic or "red tides" are the result of blooms of toxin-producing dinoflagellates (e.g., *Gonyaulax*, Figure 28.3). Toxins produced by phytoplankton accumulate in the tissues of filter-feeding organisms which have ingested the algae (e.g., some clupeid fishes, clams, mussels, and oysters). Toxic tides are not always red in color and some toxins are

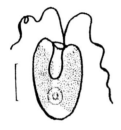

26.1 *Cryptomonas*
Phylum Cryptophyta
Scale Bar = 10 μm
(Adapted from Jahn & Jahn, 1949)

26.2 *Emiliania huxleyi*
Phylum Prymnesiophyta
Scale Bar = 1 μm
(Adapted from Green et al., 1990)

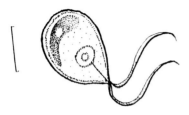

26.3 *Dunaliella*
Class Chlorophyceae
Phylum Chlorophyta
Scale Bar = 10 μm
(Adapted from Wood, 1965)

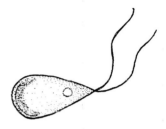

26.4 Representative Chlorophyte Flagellate
Class Ulvophyceae
Phylum Chlorophyta

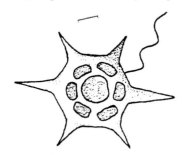

26.5 *Dictyocha*
Class Dictyochophyceae (silicoflagellates)
Phylum Chrysophyta
Scale Bar = 10 μm
(Adapted from Jahn & Jahn, 1949)

26.6 *Isochrysis*
Phylum Chrysophyta
(Adapted from Wood, 1965)

Plate 26. Representative marine flagellates

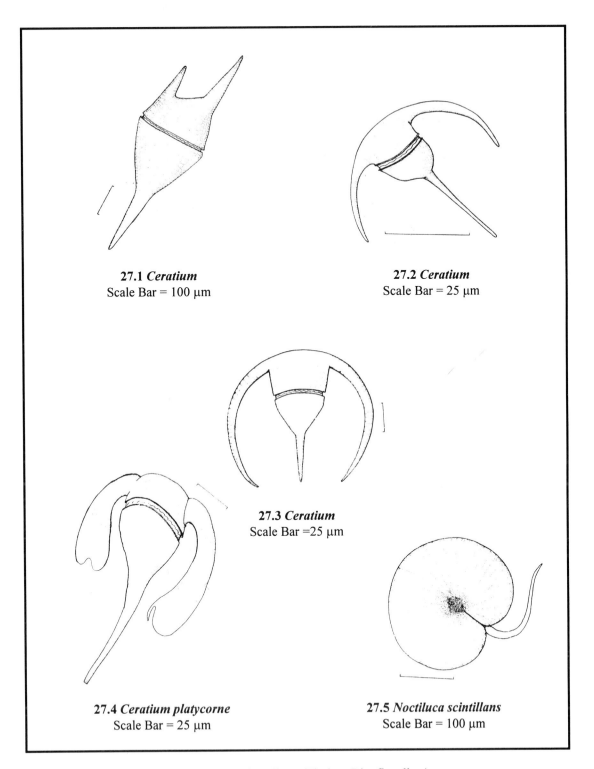

Plate 27. Dinoflagellates (Phylum Dinoflagellata)

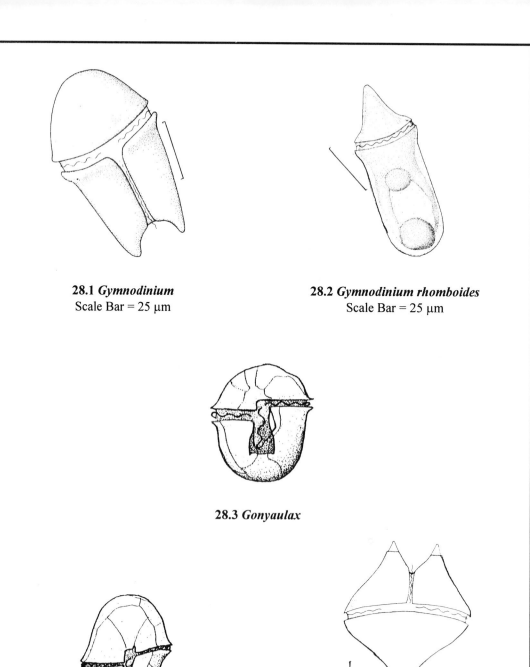

Plate 28. Dinoflagellates (Phylum Dinoflagellata)

produced by phytoplankton other than dinoflagellates (e.g., the diatom *Pseudo-nitzschia* produces demoic acid). After shellfish sequester toxins, they may be ingested by humans and cause mild to severe ailments depending upon the type of toxin. Amnesiac shellfish poisoning and paralytic shellfish poisoning are two maladies resulting from the ingestion of these toxins by humans. Gaines & Taylor (1986) provides a guide to the potentially harmful phytoplankton of the Pacific Coast of North America.

The dinoflagellate taxonomy is complex. Fensome et al. (1993) discusses how dinoflagellates are classified. Some of the more helpful works for identifying dinoflagellates include Lebour (1925) and Steidinger & Tangen (*In Press*).

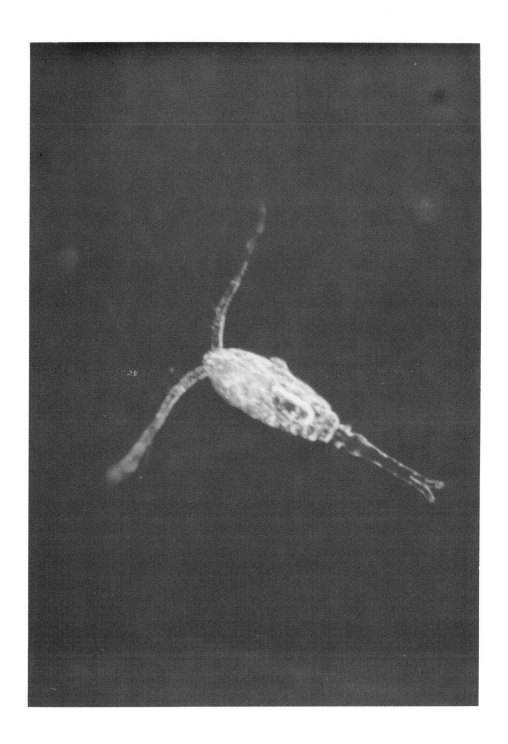

Chapter 4
Zooplankton Ecology

All of the major and most minor animal phyla are represented in the plankton. A brief investigation of coastal zooplankton offers at the same time a review of general zoology, a survey of the invertebrates, and an introduction to invertebrate embryology. The vertebrates are represented by fish eggs and young fry. Lower chordates and representatives of most of the other invertebrate phyla abound in the plankton.

Planktonic animals are either meroplanktonic or holoplanktonic. Meroplankton spend only a portion of their life-cycles, usually the larval period, in the plankton. Examples of meroplankton include crab larvae, which settle at metamorphosis to become part of the **benthic** adult population, and fish larvae, which can settle out of the plankton to become part of the **nektonic** adult community. Many meroplankton spawn eggs into the plankton where they are fertilized by a spawned male **gamete**. This **zygote** then develops through the **blastula** and **gastrula** stages to a larval form. Often larval forms are dramatically different in appearance from the familiar adults. After weeks to months in the plankton, **metamorphosis** occurs and a young adult, more like the parental form, emerges. Holoplankton, on the other hand, spend their entire life-cycles in the plankton. Examples of holoplankton include copepods and chaetognaths. Many experienced biologists and zoologists have seen these organisms only in texts and diagrams. In the plankton are found living specimens and the opportunity to watch the dramatic processes of development and metamorphosis unfold.

Specialization

Adapting to many varied requirements of planktonic existence, zooplankton often show elaborate modification. The specialization can be in the morphology, physiology, or behavior of the organism. Special adaptations for feeding, locomotion, defense and settlement are common in the plankton. Exception is the rule, as general developmental patterns give way to modifications.

Motility

Most planktonic animals have some mechanism of movement. Methods of movement include cilia, tentacles, jointed appendages, and pulsating medusa bells. Through the minor phyla are found various ciliated, free-swimming larval and adult forms. Polychaetes move using parapodia, setae, and rhythmic muscular contractions. Planktonic molluscs move by means of the ciliated **velum** of the **veliger** larva. Echinoderms also have free-swimming ciliated larval forms. The crustacea are among the fastest moving of the plankton, making effective use of jointed appendages and setae.

Feeding

Zooplankton may be carnivores, herbivores, omnivores, or detritovores. Zooplankton sometimes have specialized cells, such as the nematocysts of cnidarians, to aid in food capture. Planktonic mucous-feeders (e.g., the larvacean *Oikopleura*) cast a net or house of mucus to capture prey. Siphonophores, scyphomedusae, hydromedusae, and ctenophores extend and retract tentacles for food capture.

Some invertebrate larvae have enough yolk substance bestowed by the parent to achieve metamorphosis without feeding in the plankton. Yolk-feeding larvae with no need to capture exogenous food to achieve metamorphosis are called **lecithotrophic**. Many other larvae must ingest food in order to survive their stay in the plankton. These are said to be **planktotrophic**. These planktotrophic larvae have evolved diverse ways of capturing food. For example, some polychaetes have hard chitinous teeth or jaws for grasping. The crustacea are usually cruising, raptorial feeders and can be carnivores, herbivores, or scavengers. Many larvae, including most echinoderms and molluscs, are ciliary feeders. In these, minute cilia move water and food particles to the mouth where food is sorted and captured or rejected. Nielsen (1987) discusses the structure and function of ciliary bands in marine invertebrate larvae. Many innovative adaptations for feeding are present in the plankton.

Reproduction

Holoplankton reproduce in the plankton, while most meroplankton are the larval stage of adults which have reproduced in their nonplanktonic environment (Fig. 3). Both planktonic and nonplanktonic adults may free-spawn gametes, brood offspring for a period of time, or reproduce asexually. Examples of the latter are seen in the division of polychaete segments from the **pygidium**, protozoan fission, or the **budding** hydrozoan and bryozoan colonies. The turbellarians, scyphomedusae, and nemerteans also contain species known to reproduce asexually.

Sexual reproduction is common and diverse in the plankton. Some lower taxa form **isogametes** that fuse in zygote formation. Others possess gametes of unequal size, as the egg and sperm of the more complex life-forms. Coelenterates generally exhibit sexual reproduction in the medusa. However, in many hydroids the medusa is lacking, or reduced. Often in these cases, sexual reproduction occurs during the polyp stage. Many marine invertebrates release gametes into the water column where fertilization is not guaranteed. Common examples of this include the purple sea urchin *Strongylocentrotus purpuratus* and the green sea anemone *Anthopleura xanthogrammica*. In some benthic adults, fertilized eggs are retained as a brood for a period of time before being released into the plankton. Most barnacles brood their young in this manner. Many crustacea retain the embryos in a modified carapace or under their abdomen. For example, some mysids have a ventral brood pouch and many copepods attach zygotes to the abdomen in a single or paired ovisac. Some gastropods form exotic egg cases in which offspring develop. As in the other aspects of planktonic ecology, reproduction is achieved by unique modifications. Some modifications seem designed for practical requirements, while others to us appear random and bizarre.

Development

The development of planktonic animals may be direct or indirect. In **indirect development**, the individual develops from an embryo to a larva, then metamorphoses to an adult, as in urchins and crustaceans (Fig. 3). With **direct development**, the embryo develops into an adult with no intermediate larval form (e.g., octopus and squid).

Indirect development is often the case with meroplankton. Many marine invertebrates have a meroplanktonic larva. These larval types are often characteristic of the phylum. Some phyla have several common developmental forms. Specific representatives of a phylum may go through one, all, or some combination of these forms before metamorphosing into an adult. Common larval types for some phyla are listed in table 2 below.

PHYLUM	LARVAL TYPE
Cnidaria	Planula
	Actinula
Ctenophora	Cydippid
Platyhelminthes	Müller's
Nemertea	Pilidium
Bryozoa	Cyphonautes
	Coronate
Phoronida	Actinotroch
Annelida	Trochophore
	Metatrochophore
	Nectochaete
	Erpochaete

PHYLUM, cont'd	LARVAL TYPE, cont'd
Sipuncula	Pelagosphaera
	Trochophore
Arthropoda	Nauplius
	Metanauplius
	Cyprid
	Pre-Zoea
	Zoea
	Megalopa
Mollusca	Trochophore
	Veliger
	Pediveliger
Echinodermata	Bipinnaria
	Brachiolaria
	Echinopluteus
	Ophiopluteus
	Auricularia
	Doliolaria
	Pentacula
Hemichordata	Tornaria
Urochordata	Tadpole

Table 2. Common invertebrate larval forms by phylum.

Many plankton samples will contain an assortment of early developmental forms. Eggs, newly fertilized or unfertilized, may be found in addition to early cleavage stages (two-cell, four-cell, eight-cell, etc.).

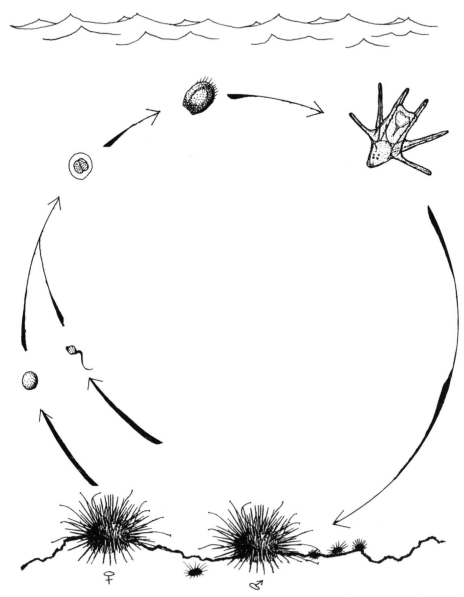

Figure 3. Representative life-cycle of a benthic marine invertebrate with planktonic larval development: the purple sea urchin *Stongylocentrotus purpuratus*.

Blastulae still within the fertilization membrane, some ciliated and slowly rotating, may also be present. Free-swimming blastulae and gastrulae, having escaped the membrane, may be observed in plankton samples. Many of these early forms, if isolated and cultured for one to a few days, will begin to exhibit early larval characteristics. Some develop rapidly, metamorphose, settle and attach, becoming post-larval or young adult forms. Using simple culture techniques, many patterns of invertebrate development can be observed.

Taxonomy and Phylogenetic Relationships

The purpose of this guide is to provide a means of identifying plankton. In order to do this, illustrations are presented which have been organized in groups of related species. For categorization, we use the widely accepted system and categories laid out by the International Commission of Zoological Nomenclature. Names assigned to various categories (e.g., in the phylum Mollusca, the name "Gastropoda" is assigned to the category of "class") can vary depending upon the opinion of the taxonomist or systematist. We use the taxonomy detailed in Brusca & Brusca (1990).

There are over thirty phyla in the kingdom Animalia. Many of these have planktonic forms which are represented in this book. Some phyla have a greater interphylum relatedness than others. At the level of phylum, groups may be considered related based upon the presence or absence of a **coelom**, whether their coelom has the same embryonic derivation, whether early cleavage is similar (e.g., **spiral** or **radial cleavage**) or if structural fate of the **blastopore** is the same (i.e., **protostomes** vs. **deuterostomes**). An example of phyla that are considered related are the deuterostomes, or those with a blastopore that becomes the anus. Deuterostome phyla include the lophophorates (Brachiopoda, Ectoprocta and Phoronida), the Chaetognatha, the Echinodermata, the Hemichordata and the Chordata. While acknowledging that controversies do exist among evolutionary biologists, we attempt to present related animal phyla consecutively, ending with the deuterostomes.

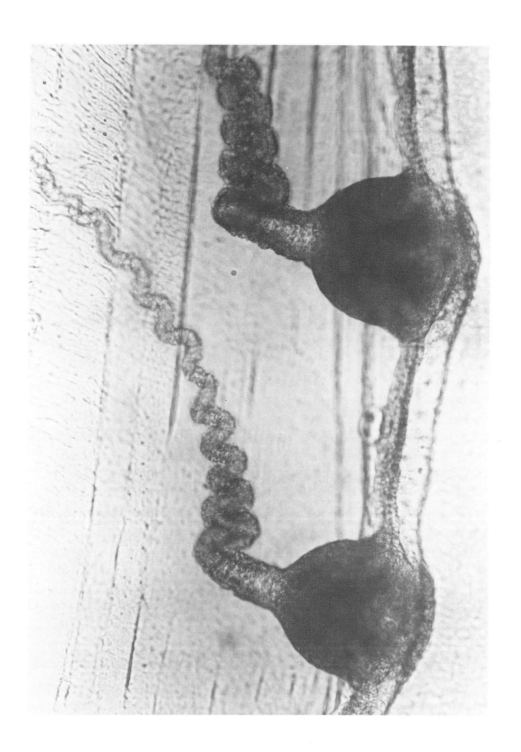

62

Chapter 5
Zooplankton Identification

Phylum Porifera

Representatives of this group of animals are not common in plankton samples. Much sponge reproduction takes place asexually by budding from the colonies. Some sponges do produce larvae that may be planktonic for hours to days. Because of the relatively short planktonic period, it is no surprise that these larvae are relatively scarce in plankton tows. However, it is possible that sponge larvae (i.e., **coeloblastulae**, **parenchymulae**, or **amphiblastulae**) may turn up in plankton samples. In general, these larvae are spherical or zeppelin-shaped and covered with cilia. Sponge **spicules** are also sometimes found in plankton samples. Spicules provide the sponge skeletal structure and, when a sponge dies, may accumulate on the sea floor. Plankton tows near the bottom or in rough water may collect spicules that have been suspended by turbulence. The shape of spicules varies according to sponge type. Some spicules are straight rods, while others are curved or tri-radiate.

Phylum Cnidaria

All cnidarian classes (Hydrozoa, Anthozoa, Scyphozoa and Cubozoa) are represented in the plankton. Cnidarians have stinging or adhesive **cnidae** which are toxic in some species. The sea wasps and box jellyfishes (class Cubozoa), among the most toxic of cnidarians, are most common in tropical seas, especially the West Pacific region near the Indian Ocean. Sea anemones, corals and octocorals (class Anthozoa) are usually planktonic only during the larval stage known as a **planula** (Plate 29). The planula is often, though not always, a zeppelin-shaped larva which has a flexible body wall covered with cilia. They swim slowly forward, rotating as they go and revolving to display internal developing **septae** and pouches (see Figure 29.1). When time has come for

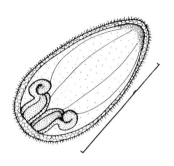 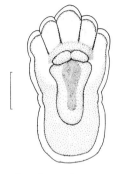 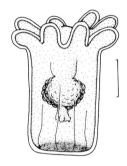

29.1 Planula, late planula and young polyp of a coral
Class Anthozoa
Scale Bars = 100 μm

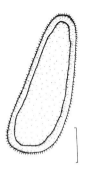 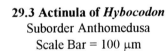

29.2 Planula of *Obelia*
Suborder Leptomedusa
Scale Bar = 25 μm

29.3 Actinula of *Hybocodon*
Suborder Anthomedusa
Scale Bar = 100 μm

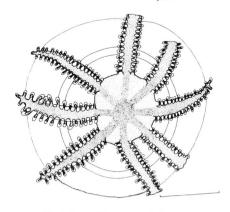 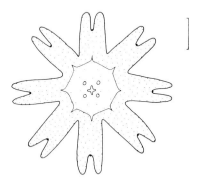

29.4 Late Planual of Octocoral
Class Anthozoa
Scale Bar = 1 mm

29.5 Ephyra of *Aurelia*
Class Scyphozoa
Scale Bar = 2.5 mm

Plate 29. Cnidaria: larvae, ephyrae and postlarval juveniles

settlement, they attach at the leading end and tentacles begin to form at the opposite, broader end (Figure 29.1).

True jellyfishes (class Scyphozoa) are conspicuously planktonic for most of their lives, though many do have benthic polypoid stages that go virtually unnoticed and are difficult to find. The prominent **medusa** stage reproduces sexually. Fertilized eggs released into the plankton develop into a planula larva, similar to that of the corals and anemones. This planula sometimes develops into an **ephyra** (Figure 29.5), which is a small medusa-like stage which develops directly into the adult. Sometimes the ephyra does not develop directly from the planula, rather the planula settles as a **scyphistoma** polyp (Figures 30.1 & 30.2) and the scyphistoma goes through a process called **strobilation** which produces the small ephyra medusa. Some of the common genera of Scyphozoa are *Aurelia* and *Chrysaora* (Figures 30.3-30.5) and *Pelagia* and *Cyanea*. These adult scyphomedusae can be quite large and usually are not captured in an average-sized plankton net. They will commonly be seen floating or swimming by the boat or dock. They are best collected with bucket or jars used to dip the animals out of the water. Before collecting large medusae, thought should be given as to where they will be kept if dissection is not to be immediate. To keep medusae alive and in good condition, large tanks with well-planned flow and drainage are needed.

Perhaps the most common cnidarians in typical coastal plankton samples are the **hydromedusae** of the class Hydrozoa. Hydromedusae have a **velum** which quickly will distinguish them from the velum-lacking scyphozoans. Hydrozoa usually have a prominent polyp stage which releases planktonic hydromedusae or **actinula** larvae (see Plates 36 & 37 for illustrations of polyps that may be caught in plankton samples if detached and swept into the water column). In those species which bear a benthic polypoid stage as well as an actinula larva, actinulae settle and metamorphose to form new young polyps. Hydromedusae often release fertilized gametes which develop into a planula which then metamorphoses into a settled polyp or a pelagic actinula.

Two orders of Hydrozoa described in this book have classic-looking medusae and are regularly encountered in the plankton. Other hydrozoan orders have species with planktonic medusae also, though

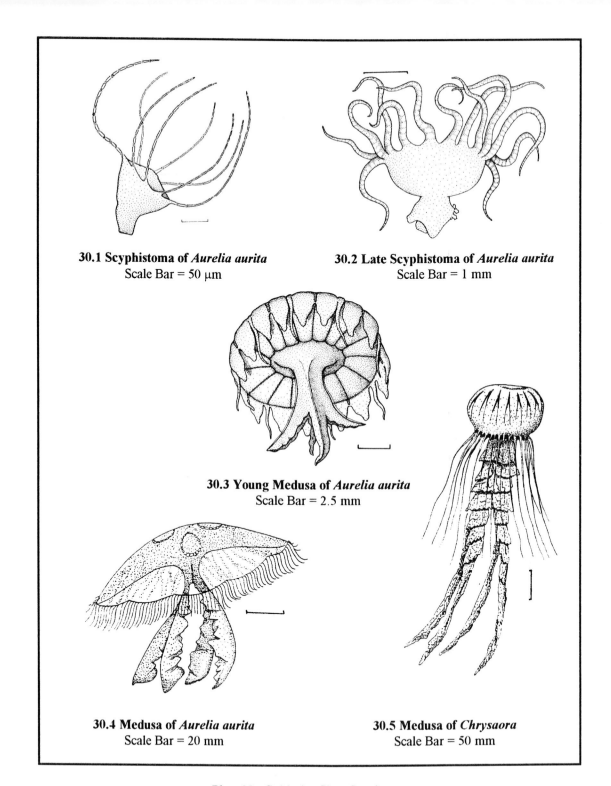

30.1 Scyphistoma of *Aurelia aurita*
Scale Bar = 50 µm

30.2 Late Scyphistoma of *Aurelia aurita*
Scale Bar = 1 mm

30.3 Young Medusa of *Aurelia aurita*
Scale Bar = 2.5 mm

30.4 Medusa of *Aurelia aurita*
Scale Bar = 20 mm

30.5 Medusa of *Chrysaora*
Scale Bar = 50 mm

Plate 30. Cnidaria: Class Scyphozoa

probably not as abundant or widely distributed geographically. One order, the Hydroida, has three suborders (Anthomedusae, Leptomedusae and Limnomedusae) whose medusae are at times abundant in plankton samples. Two more suborders of hydrozoan medusae are the Narcomedusae and the Trachymedusae (order Trachylina). Because medusae are not easily recognizable as an alternate stage of polyps, early researchers on hydromedusae sometimes unknowingly named species twice. For instance, *Phialidium hemisphericum* was named as early as 1760 and much later determined to be the medusa of the hydroid *Clytia*.

References for identifying the medusae of hydrozoans include publications from around the world. Many genera are widely distributed and these sources may be helpful for identifying hydromedusae, even if the reference is intended for a region other than that being investigated. Arai & Brinckmann-Voss (1980) and Naumov (1969) are two references used often when identifying hydromedusae. These two books include citations of many other references on hydroids and their medusae. Mills (1987) is an especially helpful guide when identifying hydromedusae in the Northeast Pacific.

When identifying medusae it is important to have a general understanding of hydromedusa anatomy. The outer and inner walls of the **bell** are the **exumbrella** and **subumbrella**. The mouth opens at the distal end of the **manubrium**, a tubular extension of the gut which hangs down from the central subumbrella. The mouth opens into the **subumbrellar cavity**, while the other end of the manubrium opens internally in the **gastrovascular cavity**. **Radial canals** extend from the center of the bell out towards the bell **margin** where they join the **ring canal**. The velum, present in most hydrozoan medusae, is a shelf-like projection of the bell margin directed inwards towards the manubrium. **Tentacles** emerge from the margin of the bell, in many cases at the junction of the radial and ring canals. **Nematocysts** are often found on tentacles, the margin of the mouth, or clustered on the outer surface of the bell. Tentacles may be solid or hollow. **Statocysts**, sensory balance organs, may or may not be present at the bell margin. The medusa swims using pulsating muscular contraction of the bell (e.g., Gladfelter, 1972).

The hyromedusa suborders and some of their distinguishing traits are described below:

Anthomedusae (order Hydroida) - An elongate bell-shaped medusa with gonads on subumbrella or manubrium. Tentacles are usually few and statocysts are absent. May or may not have ocelli. Many species lack the free medusa stage. Anthomedusae are probably the most abundant medusae in the plankton. Following these brief suborder characterizations are some descriptions of several anthomedusa genera.

Leptomedusae (order Hydroida) - Medusae generally saucer-like or flattened with gonads on subumbrella underneath the radial canals. Tentacles are usually numerous and statocysts are present. Plate 32 shows some representative leptomedusa genera including *Phialella, Eutonina* and *Phialidium* (a name once attributed to the medusae of *Clytia*). *Obelia* lacks a velum altogether and has four gonads located midway down the radial canals on the subumbrella. *Obelia* tentacles are hollow with irregular knobs of nematocysts. Statocysts are usually found at the tentacle origin or between the tentacles along the bell margin. The statocysts of most hydromedusae, including *Obelia*, are recognizable by their distinct structure of two concentric circles.

Limnomedusae (order Hydroida) - Characteristics of gonads and statocysts overlap with those of Anthomedusae and Leptomedusae. One example from the Northeast Pacific is *Proboscidactyla* (Figure 31.1) in which radial canals branch repeatedly until reaching the margin of the bell. At the junction of the radial canals and the bell margin the tentacles arise. Gonads are large, lobed and globular and surround the primary radial canal at the manubrium.

Narcomedusae (order Trachylina) - Medusa bell thin-sided with a scalloped border. Solid tentacles arise from above the bell margin. Radial canals and manubrium are lacking. The mouth opens directly into the gastrovascular cavity. The velum is broad and no radial canals are present. In the genus *Aegina*, tentacles arise above the bell margin on the exumbrella.

Trachymedusae (order Trachylina) - Hemispherically shaped, polyp reduced or lacking, velum large and the margin of the bell is not lobed as in the narcomedusae. Representative genera include *Aglantha* and *Crossota*.

Many interesting species of Anthomedusae, probably the most numerous medusae in the plankton, will be found in plankton samples. They vary in size from microscopic to a bell several cm high (e.g., *Polyorchis*, Figure 34.5). Professor Smith has seen populations of the anthomedusa *Hybocodon prolifer* reach an estimated density of several per liter in mid-spring. *Hybocodon* has a large single tentacle bulb from which one to three tentacles may extend. The other three tentacular bulbs are greatly reduced, their location marked by the ocelli. From the large tentacle, *Hybocodon* asexually buds additional medusae which break free when mature. Actinulae, collected with medusae in plankton tows, were observed to settle and metamorphose into an early polyp within 48 hours of collection.

The genus *Cladonema* (Figures 34.1 & 34.2) characteristically has five stomach lobes. From these lobes extend branching radial canals. Generally, eight of the canal branches reach the bell margin where tentacles sprout. The specimen illustrated in these Figures, taken at Bodega Bay, California, possessed nine tentacles. Evidently an extra radial canal branch extended all the way to the bell margin. The tentacles in this genus are also branched, with each branch apparently having different functions. One branch is covered with batteries of nematocysts, organized in rings, while the other is not.

Rathkea sp., Figure 33.6, possesses several unique traits. The mouth is branched with clusters of nematocysts. The eight tentacles are located are located at **radial** and **interradial** positions. Tentacles continue splitting into branches with age, until each radial tentacle has five branches and each interradial tentacle has three. The wall of the

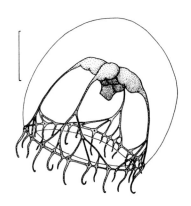

31.1 *Proboscidactyla*
Suborder Limnomedusa
Scale Bar = 1 mm

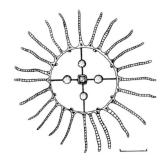

31.2 *Obelia*
Suborder Leptomedusa
Scale Bar = 100 µm

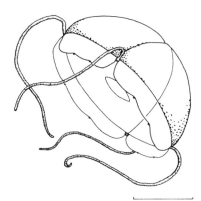

31.3 *Aegina*
Suborder Narcomedusa
Scale Bar = 1 mm

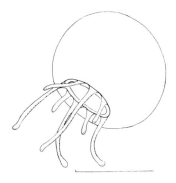

31.4 Juvenile of *Liriope tetraphylla*
Suborder Trachymedusa
Scale Bar = 25 µm

Plate 31. Cnidaria: Hydromedusae and Trachymedusae (Orders Hydroida and Trachylina)

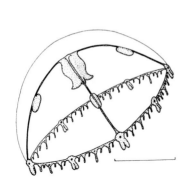

32.1 *Phialella*
Scale Bar = 1 mm

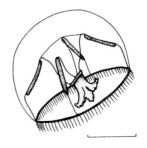

32.2 *Eutonina*
Scale Bar = 1 mm

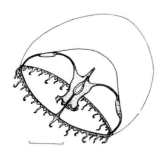

32.3 *Phialidium*
Scale Bar = 500 µm

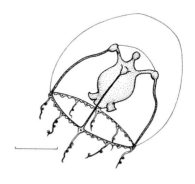

32.4 *Podocoryne*
Scale Bar = 500 µm

Plate 32. Cnidaria: Leptomedusae (Order Hydroida)

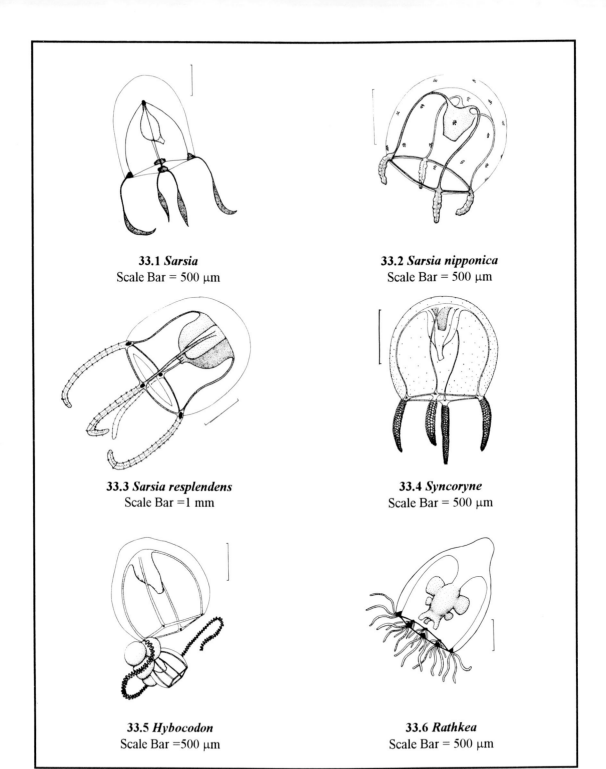

Plate 33. Cnidaria: Anthomedusae (Order Hydroida, Suborder Athecata)

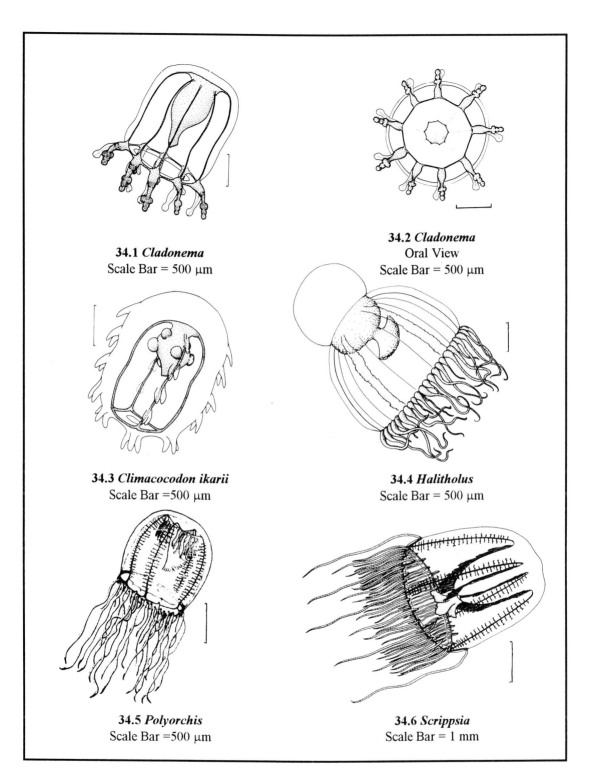

34.1 *Cladonema*
Scale Bar = 500 μm

34.2 *Cladonema*
Oral View
Scale Bar = 500 μm

34.3 *Climacocodon ikarii*
Scale Bar = 500 μm

34.4 *Halitholus*
Scale Bar = 500 μm

34.5 *Polyorchis*
Scale Bar = 500 μm

34.6 *Scrippsia*
Scale Bar = 1 mm

Plate 34. Cnidaria: Anthomedusae (Order Hydroida, Suborder Athecata)

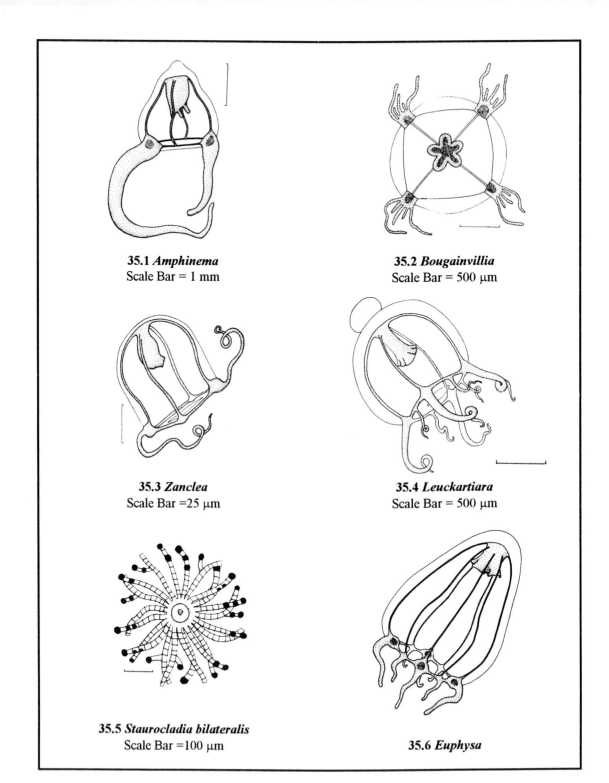

35.1 *Amphinema*
Scale Bar = 1 mm

35.2 *Bougainvillia*
Scale Bar = 500 μm

35.3 *Zanclea*
Scale Bar = 25 μm

35.4 *Leuckartiara*
Scale Bar = 500 μm

35.5 *Staurocladia bilateralis*
Scale Bar = 100 μm

35.6 *Euphysa*

Plate 35. Cnidaria: Anthomedusae (Order Hydroida, Suborder Athecata)

manubrium may have several medusa buds at various stages of maturity. Maturation is rapid and, within a day or two, new medusae will break free and swim from the subumbrellar cavity.

Bougainvillia (Figure 35.2) also has several tentacles emerging from each bulb, though the bulbs number four compared to *Rathkea*'s eight. The mouth is lobed and branched.

A single specimen of *Climacoccodon* (Figure 34.3) was taken at the wharf at Santa Cruz, California and several individuals were taken at Tomales Bay, California. Not a common genus in the East Pacific, the individual was identified using Kramp (1965, 1968). This may be a range expansion for *Climococcodon*, though, since these specimens were found, there is no record of the genus being regularly taken in the region. This medusa has tentacles arranged vertically on the exumbrellar surface directly above the radial canals. One of the specimens taken possessed outgrowths of the manubrium originally mistaken for eggs. However, in a few days actinula larvae were liberated from the outgrowths. According to Uchida (1927) these actinulae develope into a pelagic polyp, which do not settle but bud additional medusae.

Occasionally attached **polyps** of the class Hydrozoa are broken loose or fragmented and pieces may show up in plankton samples. When a branched fragment of an organism is found in the sample, chances are it is a filamentous alga or colonial animal, such as a hydroid, which has been detached from the substratum. To aid in identifying these organisms, we have included some Figures illustrating common North Pacific hydroid species. When talking about hydroid polyp stages, the suborders Anthomedusae and Leptomedusae are sometimes referred to as **gymnoblasts** and **calyptoblasts**, respectively. Calyptoblasts (Plates 36 & 37) have an extension of the chitinous exoskeleton called **perisarc**, which extends out and over the polyp as a **hydrotheca**. Gymnoblasts (Plate 37) also have a perisarc, but have no hydrotheca cup around the polyps.

Obelia, whose polyp stage has a hydrotheca, is a calyptoblast that is common on docks, boat hulls and broken in the plankton. Young leptomedusae are sometimes seen in the **gonotheca** covered **gonangium** (Figure 36.2). This animal, familiar to many biology students from charts, slides and texts, is especially delightful to observe as a living specimen.

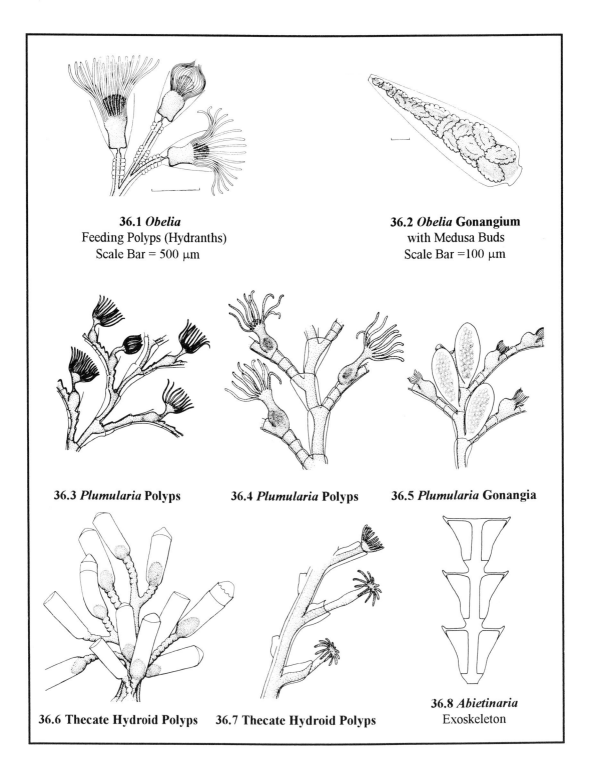

Plate 36. Cnidaria: Hydroid Polyps, Suborder Thecata (Leptomedusa)

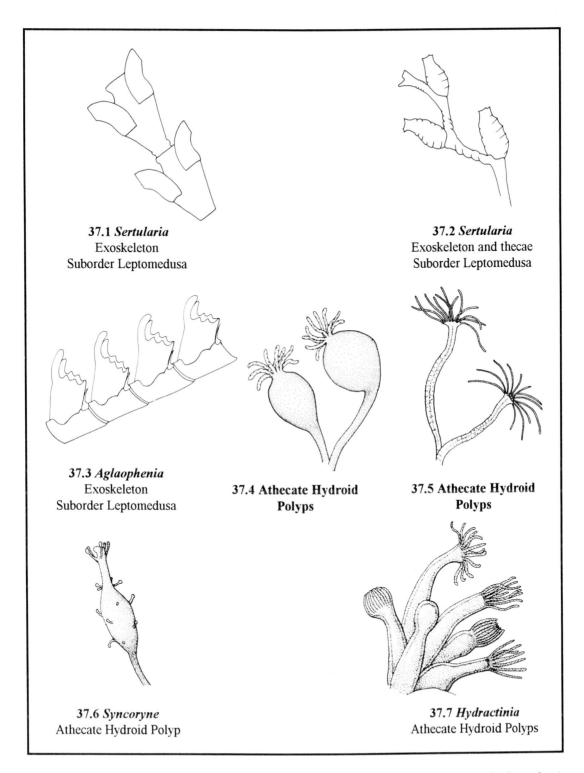

Plate 37. Cnidaria: Hydroid Polyps, Suborders Thecata (Leptomedusa) and Athecata (Anthomedusa)

The polyp moves and retracts its tentacles and food can be seen circulating in the gastrovascular cavity. Related polyps may display a wide variety of reproductive patterns. Hydroids, collected from the docks or captured in plankton tows as colony fragments, will, depending on the reproductive state of the specimen, sometimes exhibit the various reproductive strategies of the class. Some hydroids may bud medusae from the basal portion of the polyp. Some polyps lack a medusa stage (e.g., *Aglaophenia*, Figure 37.3), but possess a **corbula** where fertilized eggs are brooded until the planula emerges to settle nearby as a new polyp. *Abietinaria* and *Sertularia* (Figures 36.8 & 37.1) develop a modified sac-like gonangium. Eggs, retained on the **oogonium** are fertilized, develop to a planula larva, and then emerge and settle as a new polyp. Though the polyp is generally the asexual reproductive form, many hydroids which lack a medusa have adopted both asexual and sexual reproductive strategies in the polyp stage. When found in the plankton, these colonial hydroid fragments are worth observing, especially when reproductive structures are present and planulae or medusae may be seen emerging.

Two other orders from the class Hydrozoa that are commonly found in the plankton are the orders Siphonophora (Plates 81-84) and Chondrophora (Plate 81). Well-known examples of these orders are the Portuguese man-of-war *Physalia* (order Siphonophora) and the Purple Sailor *Velella* (order Chondrophora). Members of these two orders are generally considered to be **colonial**, with individuals specializing in function. Siphonophores are split into three suborders: the Physonectae, with a lead float and swimming bells, the Calycophorae, with no float and one or more swimming bells, and the Cystonectae, with a float and no swimming bells. The multiple swimming bells of physonect colonies are called **nectophores** (see Plate 39). A chain of nectophores with the rest of the colony trailing behind is delicate and a beautiful sight when observed *in situ*. The calycophores have a **stem** with various modified polyp groups, called **cormidia**, budding at regular intervals down its length. Each cormidium may be thought of as a subcolony of the whole. Each has a **bract** bell, a **gastrozooid** and one or more **gonophores**. Occasionally, the cormidium breaks free and lives on its own, in which case it is then called a **eudoxid**. Eudoxid gonozooid eggs can be removed and the growing

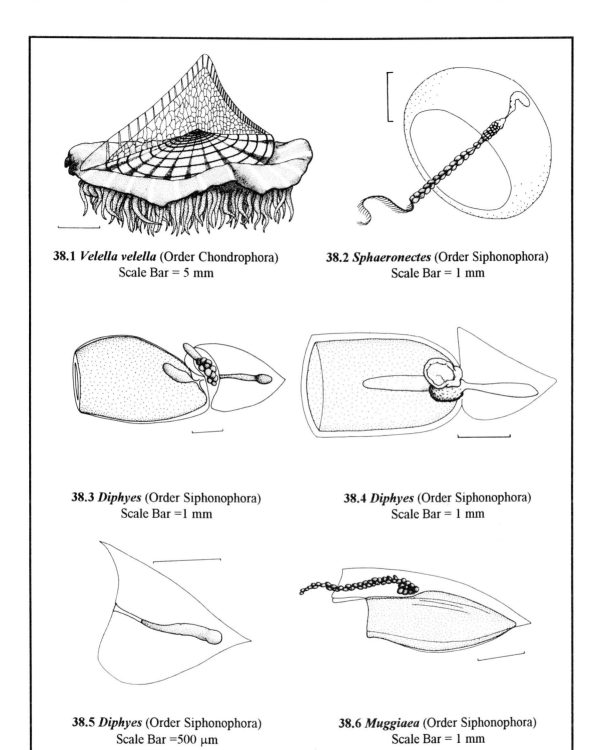

Plate 38. Cnidaria: Class Hydrozoa, Orders Chondrophora and Siphonophora

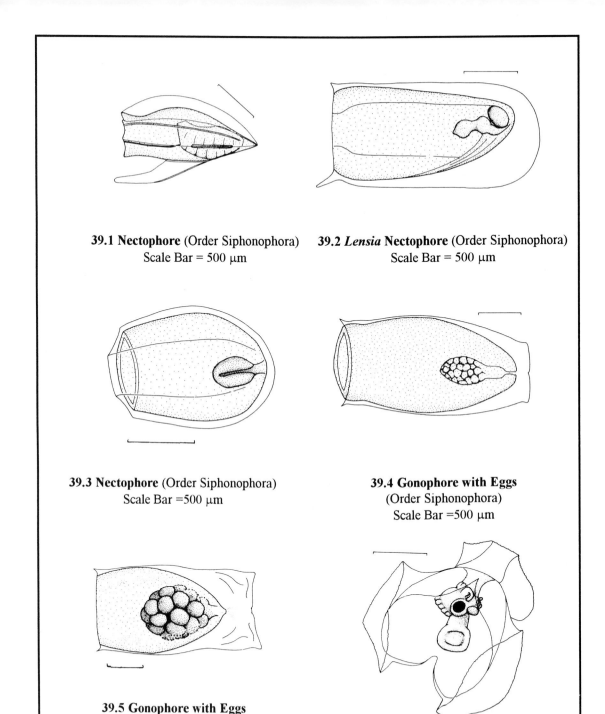

Plate 39. Cnidaria: Class Hydrozoa, Order Siphonophora. Nectophores, Gonophores and Bracts

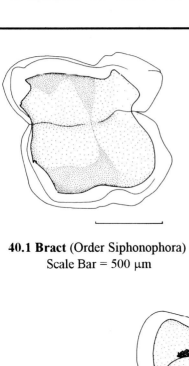

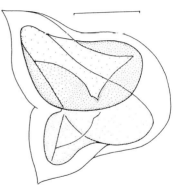

40.1 Bract (Order Siphonophora)
Scale Bar = 500 μm

40.2 Bract (Order Siphonophora)
Scale Bar = 500 μm

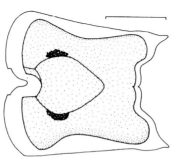

40.3 Bract (Order Siphonophora)
Scale Bar =500 μm

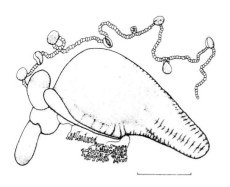

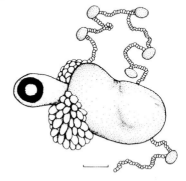

40.4 Gastrozooid (Order Siphonophora)
Scale Bar =500 μm

40.5 Gastrozooid (Order Siphonophora)
Scale Bar = 100 μm

Plate 40. Cnidaria: Class Hydrozoa, Order Siphonophora. Zooids

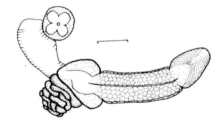

41.1 Gonozooid (Order Siphonophora)
Scale Bar = 100 μm

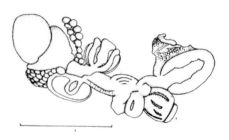

41.2 Gonozooid (Order Siphonophora)
Scale Bar = 500 μm

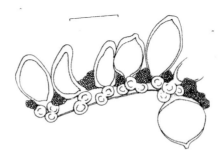

41.3 Cormidium Fragment (Order Siphonophora)
Scale Bar =100 μm

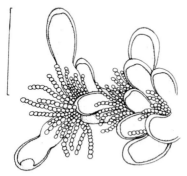

41.4 Cormidium Fragment
(Order Siphonophora)
Scale Bar =500 μm

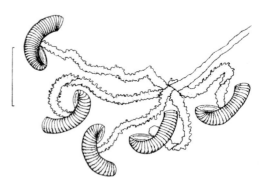

41.5 Tentacle with Tentillae
Cormidium Fragment (Order Siphonophora)
Scale Bar = 500 μm

Plate 41. Cnidaria: Class Hydrozoa, Order Siphonophora. Zooids and Cormidium Fragments

planula observed. The development of a new colony from the planula is dramatic as new nectophores develop and unfold. Plates 40 & 41 show some siphonophore zooids and fragments of cormidia.

Phylum Ctenophora

An abundant representative of this phylum, known as the comb jellies, is the sea gooseberry *Pleurobrachia* (Figure 42.4). *Pleurobrachia* begins its life as a planktonic **cydippid** larva (Figure 42.2), which then grows into the adult form. Both the adult and the larva possess eight rows of "combs", or fused cilia paddles. The beat of the cilia is rhythmic and jerking. The cilia of damaged individuals, and even fragments of individuals, may continue to beat for several hours or days provided laboratory conditions remain favorable. Torn comb jelly parts, moving with a few intact cilia, can complicate identification for the investigator who does not recognize that they are merely ctenophore fragments. The rows of Plates are **meridonal**, perpendicular to "latitudinal" lines on the spherical body of the animal. Although the beating of cilia may appear random and uncoordinated under a microscope, orientation and movement of the animal in the water is controlled and exact. Two tentacles, many times longer than the body itself, trail in graceful whorls behind the sphere to entangle and capture prey. Specialized cells, called **colloblasts**, line ctenophore tentacles and aid in capture when prey is contacted. The tentacles may be retracted into tentacular pouches, located near the sphere's "equator" within the body wall. When prey is captured, *Pleurobrachia* uses a combination of tentacle retraction and mouth reorientation to bring the quarry to the mouth.

There are a total of seven orders containing nineteen families of ctenophores. *Pleurobrachia* is a member of the order Cydippida. *Beröe* (order Beroida, Figure 42.1) is another ctenophore found occasionally in the waters of the Northeast Pacific. This comb jelly has a similar arrangement of **ctenes**, or fused cilia, as *Pleurobrachia*, but lacks the tentacles. Unlike the small mouth of *Pleurobrachia*, *Beröe* has a flat, broad mouth which enables it to engulf prey as large as or even larger than itself. *Beröe* swims mouth-first, searching for prey. If both *Pleurobrachia*

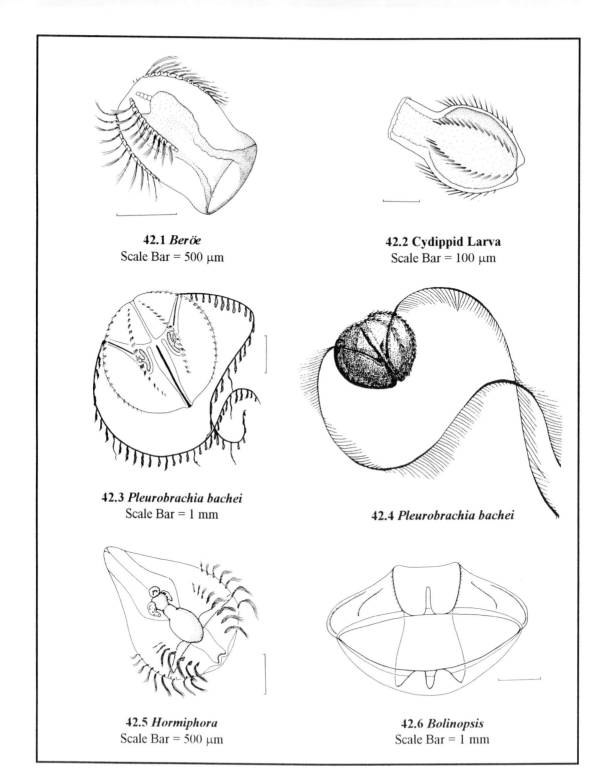

Plate 42. Ctenophora (Comb Jellies)

and *Beröe* are collected and together in the laboratory, take the opportunity to observe *Beröe* preying upon *Pleurobrachia* by placing one of each individual in the same bowl of seawater. *Beröe* can engulf a sea gooseberry as large as itself with impressive speed.

Ctenophores are **monoecious** and isolated individuals in the laboratory may release fertilized eggs which can be used for developmental studies. Young cydippid larvae will be seen within the fertilization membrane within a few days.

Additional members of this phylum are not as common in the plankton of the Northern Pacific Coast. However, other genera may be found, including *Hormiphora* (Figure 42.5) and *Bolinopsis* (Figure 42.6).

Phylum Nemertea

Occasionally the **pilidium** larva of the phylum Nemertea is present in the plankton samples. This ciliated larva (Plates 43 & 44), fascinating to observe, has a helmet-like appearance with "ear-flaps" hanging down on each side and a tuft of cilia on the crown. Pilidia glide slowly and gracefully along with the apical tuft extended, then retracting the tuft and rolling sideways to change direction. A specimen may metamorphose into a young ribbon worm (nemertean) if the pilidium is kept under favorable conditions. At times you can see the developing juvenile intestine and eyespots within the larva. Only a few pilidium larvae have been described for the Nemertea, though many of the adult forms are known. The complete life-cycle and larval identity of specific adults often are undescribed. Several pilidia are illustrated in Plate 43, but identifications are not attempted even to genus. One problem that makes planktonic ribbon worm larvae especially difficult taxonomically is that, in the past, species have been named from the larva with no knowledge of the adult form (and vice versa). An example of this is the larva known as *Pilidium recurvatum* (Figure 44.2). Which, if any, of the adult ribbon worms described in the Northeast Pacific region produce this larva is unknown. Professor Smith maintained a larval specimen of *Pilidium recurvatum*, in which could be seen the developing juvenile, in the laboratory for several days. Upon release, a tubular structure could be seen within the body

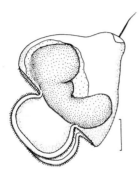

43.1 Pilidium Larva
Scale Bar = 100 μm

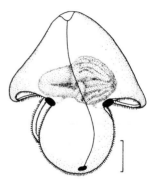

43.2 Pilidium Larva
Scale Bar = 100 μm

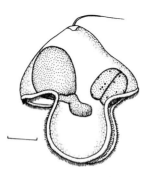

43.3 Pilidium Larva
Scale Bar = 100 μm

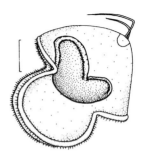

43.4 Pilidium Larva
Scale Bar = 100 μm

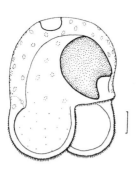

43.5 Pilidium Larva
Scale Bar = 100 μm

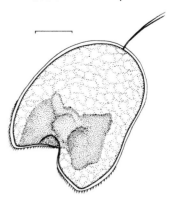

43.6 Pilidium Larva
Scale Bar = 100 μm

Plate 43. Nemertea: Unidentified Pilidium Larvae

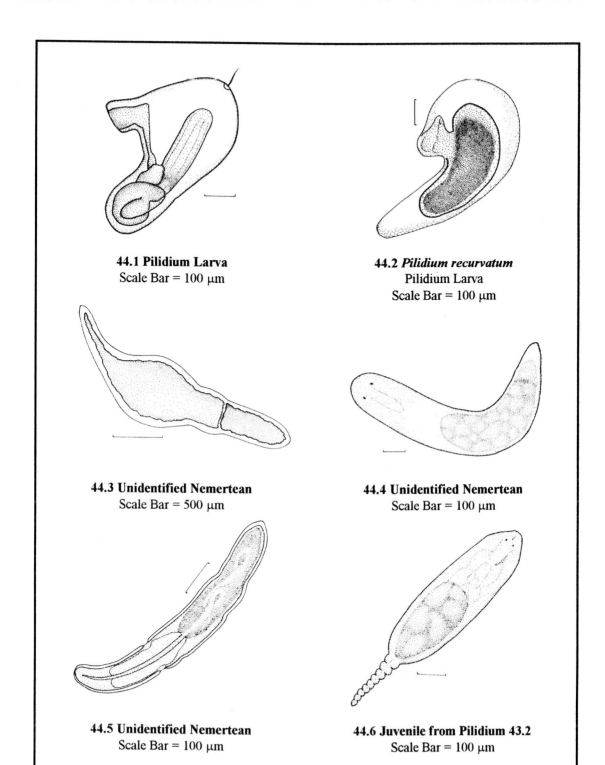

Plate 44. Nemertea: Pilidium Larvae and Juveniles

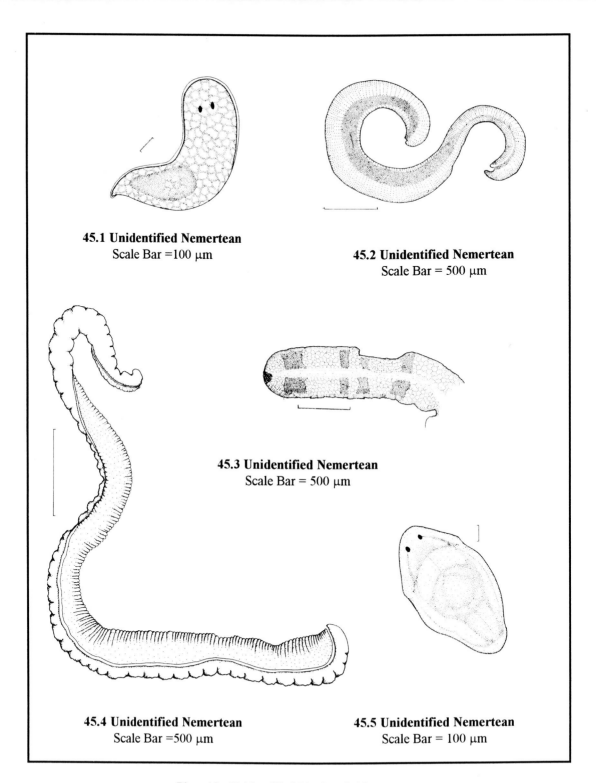

Plate 45. Unidentified Planktonic Nemerteans

cavity of the active juvenile. In a second metamorphosis this tube emerged as a later stage juvenile. Possible genera from the Northern Coast of North America, some of which undoubtedly produce most of the larvae illustrated in Plate 43 (collected in Central California), include *Tubulanus, Baseodiscus, Carinoma, Carcinonemertes, Nipponnemertes, Cryptonemertes, Nemertopsis, Ototyphlonemertes, Tetrastemma, Pelagonemertes, Micrura, Cerebratulus, Cephalothrix, Lineus, Zygonemertes, Paranemertes, Emplectonema, Amphiporus* and *Malacobdella*. Some ribbon worms are pelagic or planktonic in their adult form. Coe (1926, 1954 & 1956) has some good descriptions of pelagic nemerteans.

Phylum Platyhelminthes

Many free-living flatworms (class Turbellaria) can be found in plankton samples. Larval and adult forms living a planktonic existence, as well as those benthic forms swept into the water column by turbulence, may be captured in the plankton net as it is towed through the water. There are twelve orders of flatworms, including Polycladia, an order which includes many of the large and colorful benthic marine flatworms found in the **littoral zone** and tidepools. Polyclad flatworms have multi-branched gut, the number of gut branches being one of the criteria considered in the classification of turbellarian orders. The early larval form of the flatworm is the **Müller's larva** (Figure 46.1), recognizable by its finger-like projections and sometimes glove-like appearance. Adult individuals of *Microstoma* have been observed fragmenting into several new worms in the laboratory when removed from water.

Flukes (class Trematoda, Figures 47.4 & 47.5) are a primarily parasitic group. Though not regularly seen in the plankton and rarely in high numbers, Professor Smith has observed trematodes in the body cavities of ctenophores, medusae and chaetognaths. Usually only one parasite is found per host, though a chaetognath with two flukes and a medusa with five flukes were observed in Professor Smith's samples.

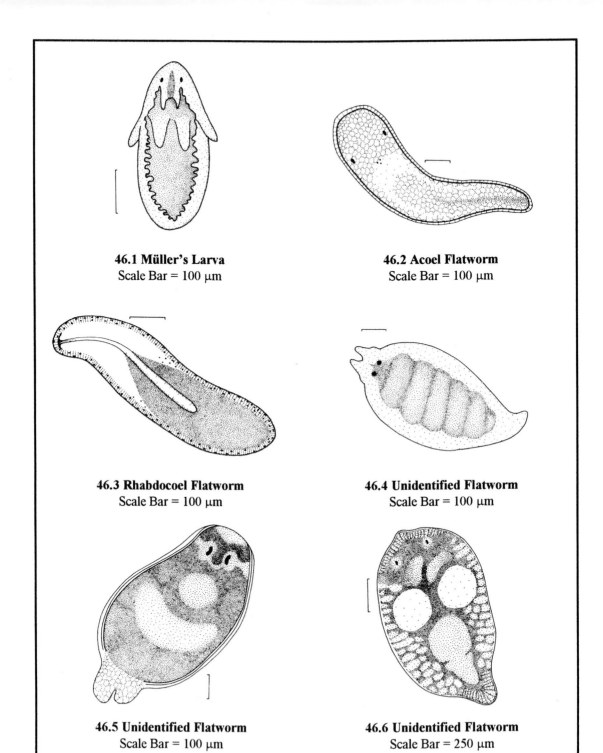

Plate 46. Turbellarian Flatworms, Larvae and Juveniles

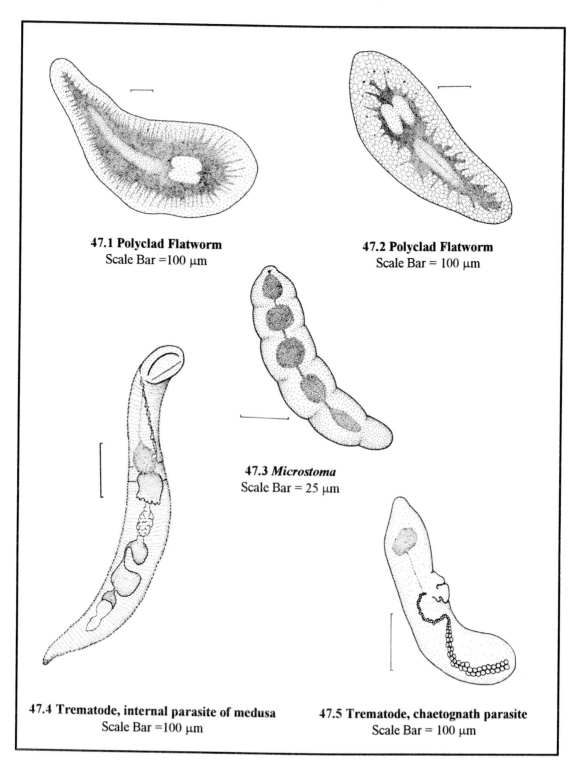

Plate 47. Platyhelminthes: Trematodes and Turbellarian Flatworms

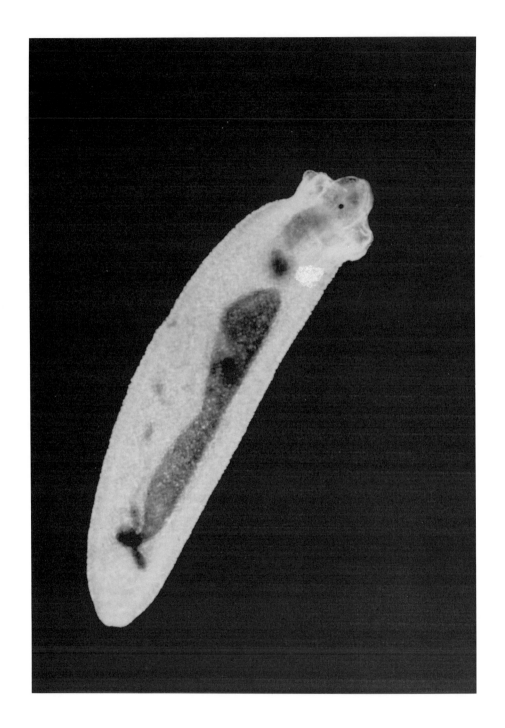

Phylum Nematoda

Nematodes, or roundworms (Figure 48.1), at times completely lacking from the plankton, can other times be the most abundant animals in the samples. A roundworm is characterized by a back and forth thrashing movement due to the flexing of **longitudinal muscles** on alternating sides of the animal. A javelin-like **fusiform** shape, pointed on both ends and slender, enables the roundworm to penetrate and burrow during locomotion and feeding.

Phylum Rotifera

Though more prominent and ecologically important in freshwater plankton, rotifers are sometimes found in marine plankton as well. A distinctive feature of the rotifer is the **corona** of cilia, found at the anterior of the animal, which appears to whirl as the cilia beat. The cilia are attached to a short region termed the head, followed by a broader trunk. The body ends with the narrow segments of the foot. Their internal anatomy can be viewed clearly through their transparent body. Some genera known to occur in coastal marine environments in the North Pacific are illustrated in Figures 48.2-48.4.

The Phyla Sipuncula and Echiura

This phylum is represented along the Pacific Coast of North America by several genera, including *Golfingia*, *Phascolosoma* and *Themiste*. The phylum's planktonic larval forms are the **trochophore** and the **pelagosphaera** (Plate 49). Trochophores of sipunculans are similar in appearance to mollusc trochophores. Pelagosphaera larvae are up to three millimeters in length and often visible to the naked eye. Using the heavily ciliated metatroch, where a strong wave of large cilia almost constantly beats, they cruise through the water anterior end first. The cephalic region, anterior of the metatroch, bears eyespots and can be retracted into the body. When the head is retracted, the larva ceases swimming and sinks. This behavior, exhibited when the animal is disturbed, may be a defensive

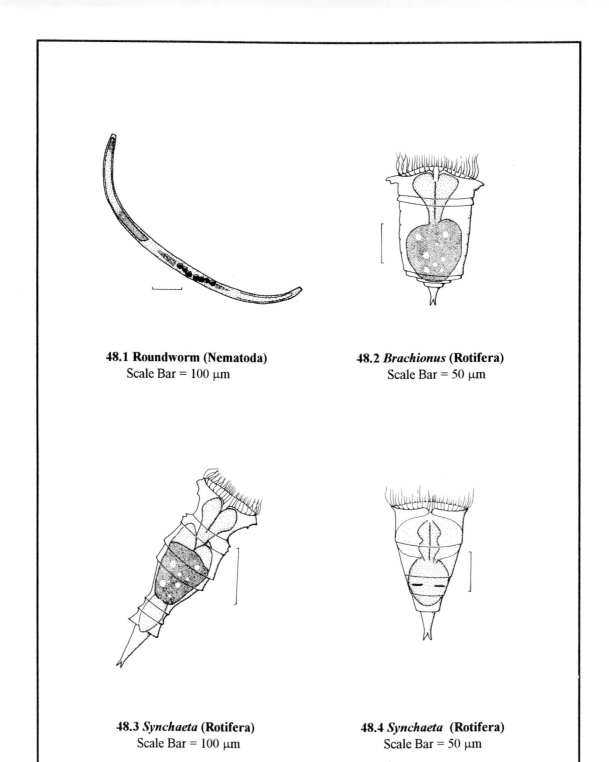

Plate 48. Phylum Nematoda and Phylum Rotifera

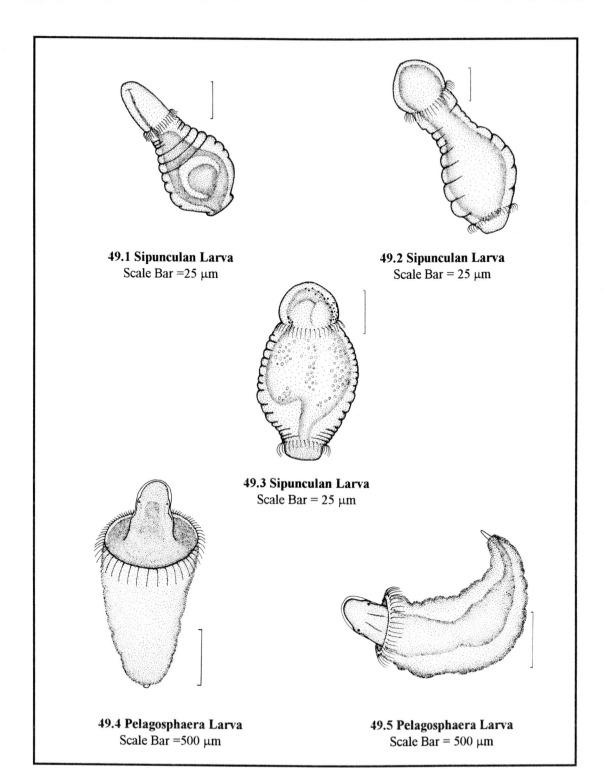

Plate 49. Sipuncula: Trochophore and Pelagosphaera Larvae

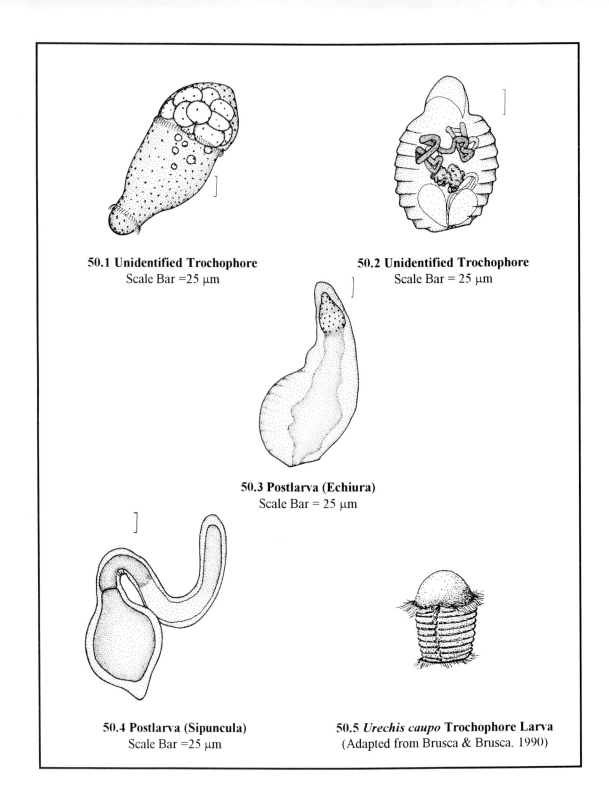

Plate 50. Trochophore Larvae of Unknown Affinity, Sipunculan and Echiuran Developmental Stages

response. As metamorphosis approaches, the pelagosphaera may do "headstands" and occasionally take the most posterior projection, or tail, in its mouth as it rolls in a tight circle. Upon metamorphosis the larval features are lost and the form of an adult sipunculan emerges. Plate 50 includes some pelagosphaera-like and trochophore-like larvae thought to be sipunculans or echiurans. Echiurans have a planktonic trochophore larva. The trochophore of *Urechis caupo* (Figure 50.5), the fat innkeeper worm, looks much like a sipunculan trochophore, but without the pronounced **apical tuft**.

Rice (1967) and Rice (1975a) provide illustrations and photographs of the trochophores and pelagosphaerae of the sipunculan genera *Phascolosoma, Golfingia* and *Themiste*. Rice (1975b) and Rice (1976) give reviews of development in the phylum Sipuncula. Jägersten (1963) discusses morphology and behavior of sipunculan larvae.

Phylum Annelida

The phylum Annelida has many strikingly diversified forms. Annelids are divided into three classes: the Polychaeta (primarily marine), the Oligochaeta, and the Hirudinoidea (the leeches). Most biology students have been introduced to this phylum by the terrestrial earthworm *Lumbricus terrestris* (class Oligochaeta), and some students may be familiar with the **infaunal** marine polychaetes *Neanthes, Nephtys* or *Nereis*. A greater appreciation of the complexity and diversity of this group can be gained by the use of plankton in classroom study. The planktonic realm's many polychaete annelids display a variety of sizes, colors, forms, and developmental modes.

It is not surprising to find twenty or more families of polychaetes in plankton samples taken over several months in coastal waters. Planktonic polychaetes, like those that live on or in the benthos, vary in structure, size and activity. Adult polychaetes sometimes use cilia in specialized structures for feeding (e.g., *Eudistylia*), locomotion, respiration, etc. There are also many different patterns of reproduction and development. In general, the polychaetes are **dioecious**, having separate sexes, while the oligochaetes are, as a rule, **hermaphroditic**. In marine polychaetes,

gonads often do not exist and gametes are produced by cell division from the wall of the **coelom**. In some species, the walls produce bulges in the segments which erupt and expel gametes into the surrounding seawater. In the pelagic *Tomopteris* (Figure 52.7), eggs are stored in branches of the coelom which extend out into the **parapodia**. In others (e.g., *Exogone*, Figure 52.6) eggs are attached to the outer body wall to begin development. *Autolytus* (Figures 52.3-52.5) forms a **brood** mass of fertilized eggs which develop through the trochophore larval stage and are released to begin a planktonic life when the larvae are three to five segments in length. Most marine polychaetes produce a planktonic trochophore larva. As this trochophore grows, it adds segments and comes to be known as a **polytroch** or **metatrochophore**. Nereid (family Nereidae) eggs can often be found settling in the bottom of plankton samples and appear to contain several large yolk globules within the membrane. Some planktonic developmental stages of nereids are illustrated in Plate 54. Figure 54.2 is an example of a nereid **nectochaete** larva. Other polychaetes retain their eggs within the egg membrane through late larval development. Figure 51.4 was drawn from a photograph taken immediately after release of the animal from the egg membrane. Early release is the rule with spionids (family Spionidae), but there are some exceptions (Figure 53.1). Professor Smith observed this larva developing within the egg membrane with its long **chaetae** piercing and protruding through it. As the larva grows, the membrane draws ever closer around the larva until the larva is encased in the tightly fitting egg membrane. In general, polychaete trochophore larvae possess a large ciliary band just anterior to their widest girth called a **prototroch**. On the anterior crown of the trochophore is often found an apical tuft of cilia for sensory perception. Often a posterior ring of cilia around the trochophore, called a **telotroch**, is also present. Sometimes a ventral band of cilia, the **neurotroch**, extends between the prototroch and the telotroch. This theme and variations on it can be seen in the myriad trochophores illustrated in Plate 51.

    Some syllids (family Syllidae), nereids (family Nereidae) and eunicids (family Eunicidae) can produce a sexually reproductive individual with a posterior **epitoke** containing gonads. With this unusual

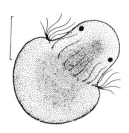

51.1 Polychaete Trochophore
Scale Bar = 100 μm

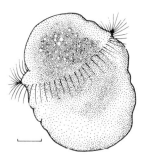

51.2 Polychaete Trochophore
Scale Bar =100 μm

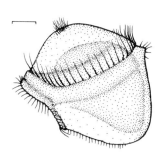

51.3 Polychaete Trochophore
Scale Bar = 100 μm

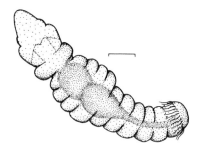

51.4 Polychaete Trochophore
Scale Bar = 100 μm

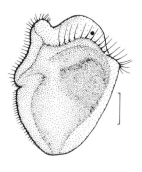

51.5 Polychaete Trochophore
Scale Bar = 100 μm

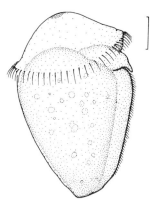

51.6 Polychaete Trochophore
Scale Bar =100 μm

Plate 51. Polychaete Annelids: Early Trochophore Larvae

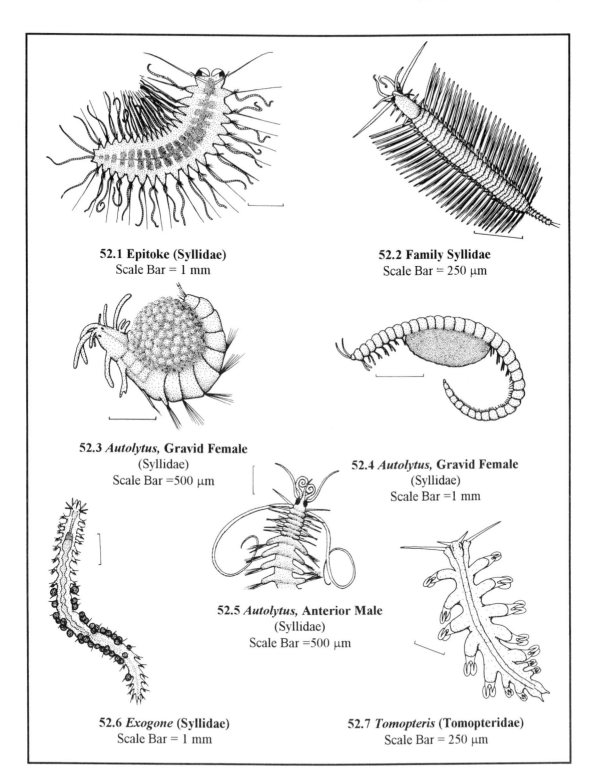

Plate 52. Polychaete Annelids: Families Syllidae and Tomopteridae

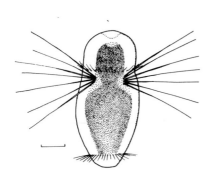

53.1 Trochophore Larva (Spionidae)
Scale Bar = 100 μm

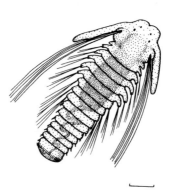

53.2 Metatrochophore Larva (Spionidae)
Scale Bar = 100 μm

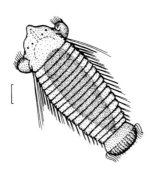

53.3 Metatrochophore Larva (Spionidae)
Scale Bar = 100 μm

53.4 Metatrochophore Larva (Spionidae)
Scale Bar = 100 μm

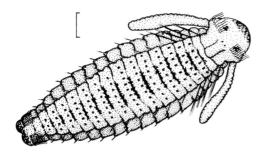

53.5 Metatrochophore Larva (Spionidae)
Scale Bar = 250 μm

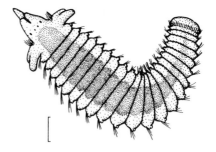

53.6 Metatrochophore Larva (Spionidae)
Scale Bar = 250 μm

Plate 53. Polychaete Annelids: Family Spionidae

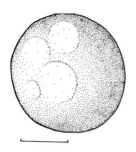

54.1 Polychaete Egg (Nereidae)
Scale Bar = 50 μm

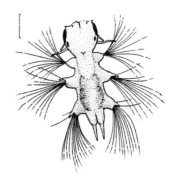

54.2 Nectochaete Larva (Nereidae)
Scale Bar = 50 μm

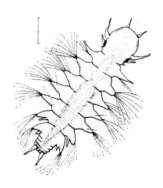

54.3 Late Nectochaete Larva (Nereidae)
Scale Bar = 250 μm

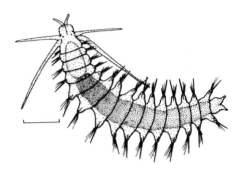

54.4 *Platynereis* (Nereidae)
Scale Bar = 500 μm

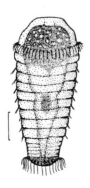

54.5 *Nephtys* Metatrochophore (Nereidae)
Scale Bar = 100 μm

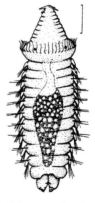

54.6 *Glycinde* Metatrochophore (Goniadidae)
Scale Bar = 100 μm

Plate 54. Polychaete Annelids: Families Nereidae and Goniadidae

reproductive modification, the anterior ends of the individuals are often atoke, or lacking the gonads. Epitokes (e.g., Figure 52.1) can break free and become a free-swimming reproductive structure. In some epitokes, eyes and tentacles develop in the anterior end in imitation of the adult worm.

In most polychaetes, a fairly dramatic metamorphosis takes place between the larval and juvenile stages. The adult identity of a planktonic polychaete larva can seldom be determined by relying upon morphological similarities of the larva to the adult. The larvae of *Chaetopterus* (Figures 55.1-55.3), *Owenia* (Figure 57.4) and *Polygordius* (Figure 57.1) are a few of the unusual planktonic forms that undergo metamorphosis to become a benthic adult. The *Owenia* larva shown in Figure 57.4 is a unique larval form called a **mitraria**. The cataclysmic metamorphosis of a mitraria larva to an adult annelid is described by Wilson (1932). *Polygordius*, an archiannelid of the order Polygordiida, grows via increased segmentation, finally dropping from the hood (Figure 57.2) to elongate. Some capitellids (order Capitellida) elongate until a certain number of segments are present and then metamorphose to the tube-dwelling post-larval form. The larva of *Eupolymnia* (Figure 57.6) requires the presence of a suitable substratum before settling to build the familiar honeycombed tubes of the adult form. As with other planktonic invertebrate larvae, much can be learned from isolating individuals and maintaining them in the laboratory to observe behavior and metamorphosis.

Most early polychaete larvae retain yolk granules from the egg as a food substance for growth and development. As the yolk is gradually used up, the need for supplementary food increases. Cultures of diatoms and unicellular marine algae are often used to provide the necessary nutrition to planktotrophic larvae being maintained in the laboratory. Some phyllodocids (Plate 56, family Phyllodocidae) are predaceous carnivores and may even eat other larvae.

When developing from a trochophore, the animal grows by adding new segments posteriorly. The terminal segment is known as the **pygidium**. The pygidium can bud new segments, which in some polychaetes can break free as an asexually reproduced individual. Thus, the foremost segments are older and those immediately preceding the

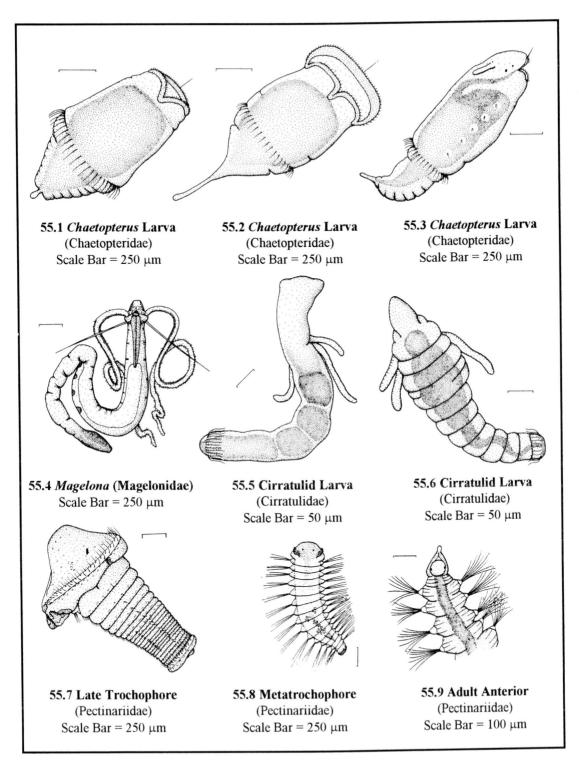

Plate 55. Polychaete Annelids: Families Chaetopteridae, Magelonidae, Cirratulidae and Pectinariidae

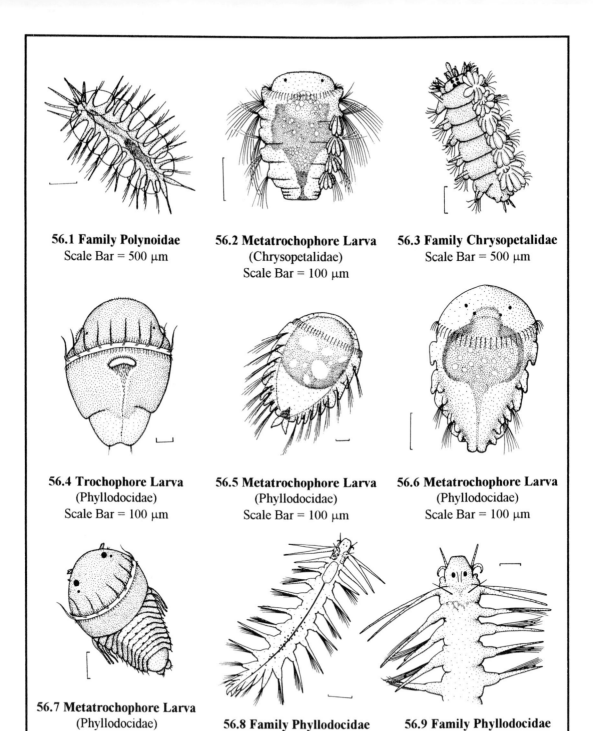

Plate 56. Polychaete Annelids: Families Phyllodocidae, Polynoidae and Chrysopetalidae

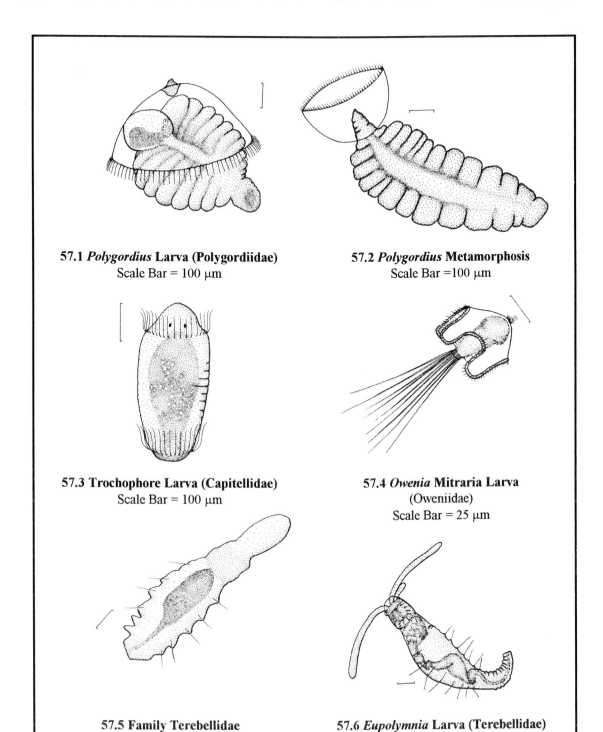

Plate 57. Polychaete Annelids: Families Polygordiidae, Oweniidae, Capitellidae and Terebellidae

58.1 Early Metatrochophore (Sabellaridae)
Scale Bar = 50 μm

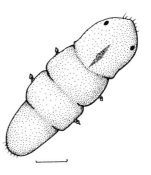

58.2 Erpochaete Larva (Lumbrineridae)
Scale Bar = 50 μm

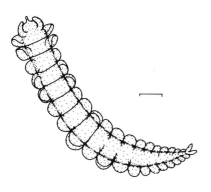

58.3 *Typhloscolex* (Typhloscolecidae)
Scale Bar = 500 μm

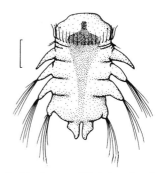

58.4 *Ophiodromus* Metatrochophore
(Hesionidae)
Scale Bar = 500 μm

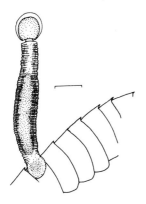

58.5 Leech on Carapace (Class Hirudinea)
Scale Bar = 250 μm

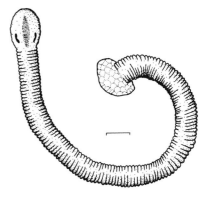

58.6 Leech (Class Hirudinea)
Scale Bar = 250 μm

Plate 58. Annelida: Hirudineans and Representatives of Four Polychaete Families

pygidium are the newest. This chain-like development is termed **teloblastic** growth.

The regenerative capabilities of annelids are impressive. One specimen of a post-larval polychaete was taken in Tomales Bay, California. Possibly due to a predator, the pygidium and several posterior segments were lacking. After observation in the laboratory for two weeks, the pygidium (regenerated after four days) and six to eight segments (regenerated at a rate of one per day after the appearance of the new pygidium) were present.

Locomotion in polychaetes is accomplished by swimming, paddling and crawling with parapodia and setae, though early developmental forms show effective use of cilia for locomotion. Specialized swimming, burrowing, feeding and sensory structures (e.g., parapodia, chaetae, tentacles and palps) display a wide range of form and utility. Many tube-dwellers can be induced to begin tube formation if placed in a culture dish and provided with a small amount of clean sand. Some polychaetes produce a gelatinous sheath, upon which sand grains are placed. Other newly metamorphosed tube-dwellers fashion their tubes from calcareous or chitinous material.

Representatives of the class Hirudinoidea (=Hirudinea, Hirudinida), the leeches, are seldom found in the plankton. However, some of these parasitic forms are occasionally seen attached to the carapace of a mysid shrimp, an isopod or some other crustacean. They possess segments, though not as visually distinct as polychaete segments, and ventral suckers (Figures 58.5 & 58.6).

The majority of planktonic polychaetes are larval forms. However, some planktonic adult forms exist and these may be identified using Kozloff (1987). Larval polychaetes can usually be identified to family without too much difficulty. One of the more comprehensive collections of polychaete larval identifications is by Bhaud & Cazaux (1987). Other polychaete larval guides (Korn, 1960; Lacalli, 1980) are prepared for specific regions (i.e., not the North Pacific), but include most of the pertinent families (and often genera) and can be quite useful. Descriptions of development from representatives of the following families are indicated by the accompanying references: the Onuphidae (Blake, 1975a), the Magelonidae (Wilson,

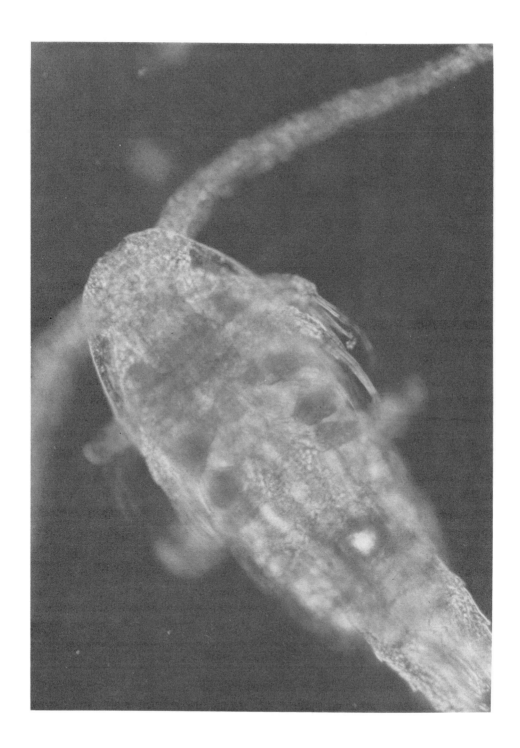

1982), the Spionidae (Levin, 1984) and the Sabellariidae (Eckelbarger 1976, 1977). Larval development of eighteen California coastal species from ten families is described in Blake (1975b). Hannerz (1961) provides an illustrated key to the Spionidae, Disomidae and Poecilochaetidae. Ciliary structure and function in *Owenia's* mitraria larva is described by Emlet & Strathmann (1994) in a chapter which includes several photographs of the mitraria larva.

Phylum Arthropoda

The planktonic arthropods are primarily crustaceans. There are three classes of noncrustacean arthropods that may occasionally turn up in plankton samples close to the shore: class Pycnogonida (subphylum Chelicerata), class Arachnida (subphylum Chelicerata) and class Insecta (subphylum Uniramia). These noncrustaceans are often incidentally planktonic, rather than being adapted specifically and exclusively for planktonic life. The salt water mites (class Arachnida, Plate 59) are most often found clinging to bits of debris and algae collected close to the intertidal zone. Adult sea spiders (class Pycnogonida), often parasitic or commensal with sea anemones, hydroids or algae, are sometimes swept from benthic subtidal or intertidal habitats where they are collected in plankton samples. Their unique larval form, the protonymphon, is usually **symbiotic** with larger animals from other phyla. Some adult and larval sea spiders may be captured along with pelagic cnidarians or other planktonic hosts. *Nymphon* (Figure 59.1) is regularly found in near-shore plankton samples from the Central California coast, while *Achelia* and *Pycnogonum* are only rarely captured incidentally in plankton samples. Mites, seaspiders and some supralittoral insects are often washed into the water column by wave action and currents. Insects are not included here because of their unmanageable global variety and natural terrestrial habitat.

Subphylum Crustacea (Phylum Arthropoda)

This diverse group, with all its classes, subclasses, orders and suborders, accounts for a large portion of most plankton samples. All

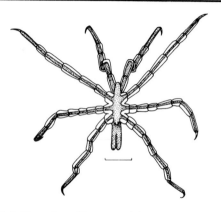

59.1 *Nymphon* (Pycnogonida, Nymphonidae)
Scale Bar =500 μm

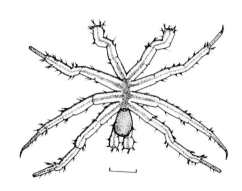

59.2 *Pycnogonum* (Pycnogonida, Pycnogonidae)
Scale Bar = 500 μm

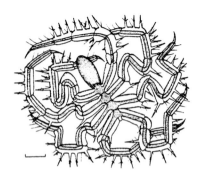

59.3 *Achelia* (Pycnogonida, Ammotheidae)
Scale Bar =500 μm

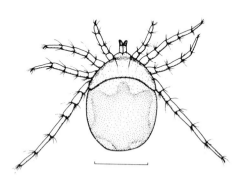

59.4 Salt Water Mite (Arachnida, Halacaridae)
Scale Bar =250 μm

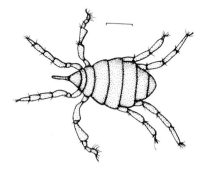

59.5 Salt Water Mite (Arachnida, Halacaridae)
Scale Bar =250 μm

Plate 59. Arthropods: Class Pycnogonida and Class Arachnida

members of this phylum, larvae and adults alike, possess the phylum's distinguishing jointed appendages. In addition to planktonic copepods and their larvae, the larvae of many benthic and pelagic crustaceans abound in the plankton. Like copepods, most of these other crustaceans hatch out as a **nauplius** larva (various examples, Plate 60), with three pair of jointed appendages. Nauplii usually have a single compound eye, known as a **naupliar eye**, which becomes modified in later larval stages. Many go through more than one, sometimes several, larval molts before metamorphosing into an adult. As nauplii and other specialized crustacean larvae, such as the zoea larva of the decapoda, pass through subsequent molts and/or morphology shifts on their journey to metamorphosis, the larval stage is numbered according to how many molts or stages the larva has passed through (e.g., nauplius I and nauplius II for before and after the first naupliar molt, respectively).

Modifications on the crustacean body theme vary from simple to extreme. Still, there is enough consistency in the end-product (the adult crustacean) that obvious **homologous** segments and appendages exist. The terminal body segment in crustaceans is known as the **telson**. In many groups it is flanked by a pair of **biramous uropods**. This segment and these appendages are collectively called a **tail-fan**. In some taxa, certain body segments are fused when compared to the primitive crustaceans. This process of segment fusion and specialization is known as **tagmosis**. The segments in the head-region are often fused and house the small crustacean brain. The tendency in animals for the head to bear the most critical components of the nervous system is called **cephalization**. Some groups (e.g., crabs) turn the abdomen underneath the fused **cephalothorax**. Appendages may be highly modified for a specific function in many of the more evolved groups. We discuss the major crustacean groups found in the plankton, their distinctive features and life history. Crustacean larval forms are among the most abundant animals in the plankton. Their diversity can be overwhelming and it is recommended that the investigator seek an understanding of general form and taxonomy of the adult and larval crustaceans. An overview of larval morphology and diversity in the crustaceans is given by Williamson (1982).

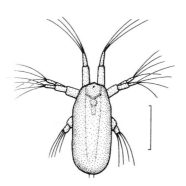
60.1 Unidentified Nauplius Larva (Crustacea)
Scale Bar = 100 μm

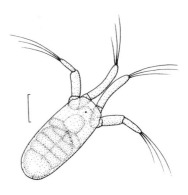
60.2 Unidentified Nauplius Larva (Crustacea)
Scale Bar = 50 μm

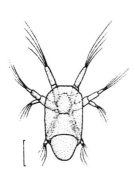
60.3 Unidentified Nauplius Larva (Crustacea)
Scale Bar = 100 μm

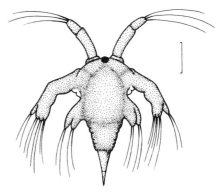
60.4 Unidentified Nauplius Larva (Crustacea)
Scale Bar = 250 μm

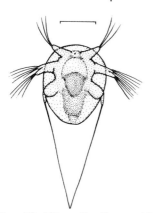
60.5 Unidentified Nauplius Larva (Crustacea)
Scale Bar = 50 μm

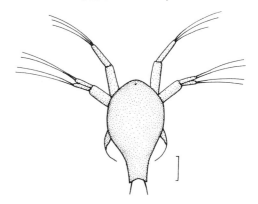
60.6 Unidentified Nauplius Larva (Crustacea)
Scale Bar = 50 μm

Plate 60. Crustacean Nauplius Larvae

Subclass Cirripedia (Barnacles)

The barnacles (class Maxillopoda) are represented in the plankton primarily by two larval forms. The first, as in many other Crustacea, is the shield-shaped nauplius (Plate 61). This stage is followed by the settling stage known as a **cyprid**. The cyprid of both acorn barnacles and gooseneck barnacles has a similar body form to the latter. When settling, the cyprid, having found a suitable spot by chemical or some other cue, cements itself to the substratum and begins its life as a **sessile** organism. Newly settled barnacles are essentially glued to the rock on their backs with appendages, highly modified for passive and active filter-feeding, projecting outward from the rock. Barnacle molts are also found in the plankton. These molts (e.g., Figure 61.5) can be quite common, especially in samples taken near docks, intertidal rocks or anywhere that settled adult barnacles are abundant.

Extensive keys to several common genera of East Coast barnacle larvae (Lang, 1977) are helpful in many regions, including the North Pacific, as they include many genera global in distribution. Detailed illustrations of spines and setae are usually needed to identify barnacle larvae to species. Several larval descriptions of Pacific Coast species of the genus *Balanus* are available (e.g., Barnes & Barnes, 1959; Branscomb & Vedder, 1982; Brown & Roughgarden, 1985). Larval descriptions of representatives of *Tetraclita* and *Megabalanus* are described in Miller & Roughgarden (1994). Larval development of a common West Coast rocky shore gooseneck barnacle, *Pollicipes polymerus*, is shown in photographs and illustrations in Lewis (1975).

Subclass Ostracoda

Ostracods (class Maxillopoda) have a large bivalve carapace which is bean-shaped and has earned ostracods the nick-name "bean clams". Indeed, they are often confused with clams when their appendages are folded within the valves. They are also sometimes confused with the barnacle cyprid larvae. This is somewhat understandable, though one can see that ostracod appendages (Figure 62.6) are distinct from barnacle cyprid appendages (Figure 61.4). Ostracods are found in both freshwater and marine environments. The compound eye is usually not apparent and

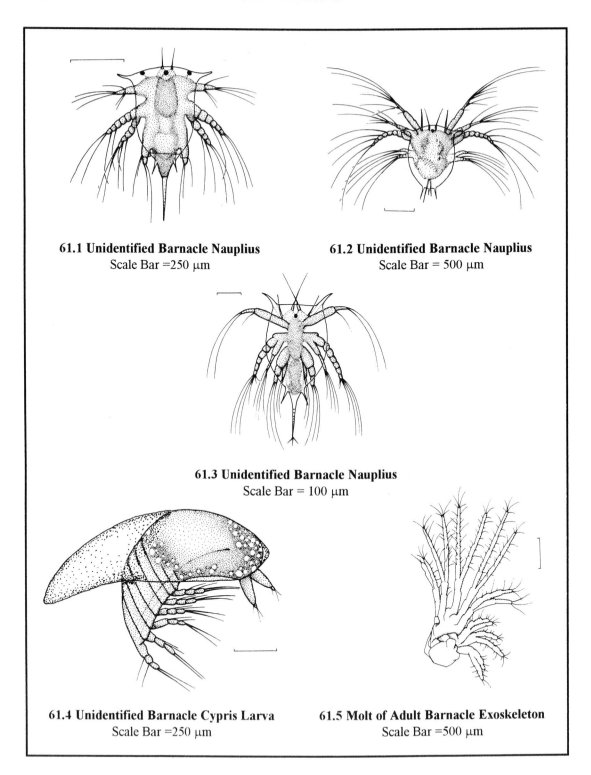

Plate 61. Crustacea: Barnacle larvae and molts (Subclass Cirripedia)

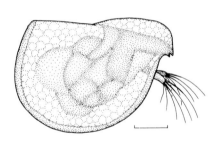
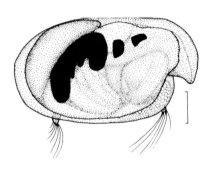

62.1 *Conchoecia* **(Order Halocyprida)**
Scale Bar = 100 μm

62.2 *Conchoecia* **(Order Halocyprida)**
Scale Bar = 100 μm

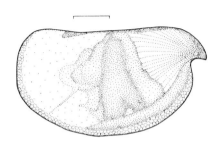
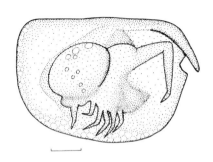

62.3 *Conchoecia* **(Order Halocyprida)**
Scale Bar = 100 μm

62.4 Unidentified Ostracod
Scale Bar =100 μm

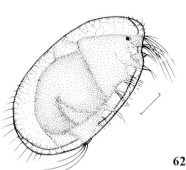
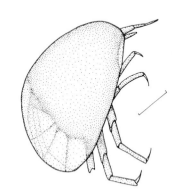

62.5 Unidentified Ostracod
Scale Bar =100 μm

62.6 *Cypridina* **(Order Cypridinida)**
Scale Bar =100 μm

62.7 Unidentified Ostracod
Scale Bar =100 μm

Plate 62. Crustacea: Ostracoda (Ostracods)

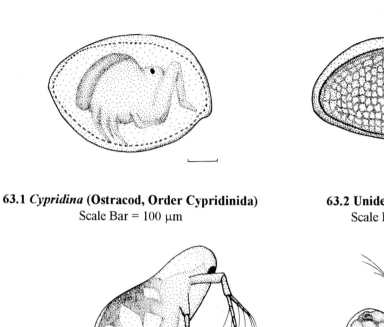

63.1 *Cypridina* **(Ostracod, Order Cypridinida)**
Scale Bar = 100 µm

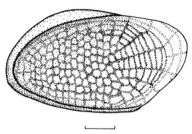

63.2 Unidentified Ostracod
Scale Bar = 100 µm

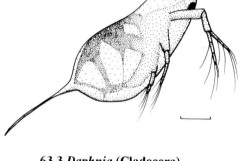

63.3 *Daphnia* **(Cladocera)**
Scale Bar = 100 µm

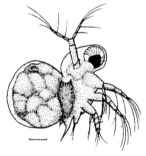

63.4 *Podon***, Gravid Female (Cladocera)**
Scale Bar = 100 µm

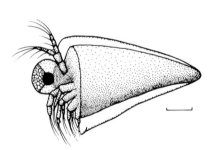

63.5 *Evadne***, Male (Cladocera)**
Scale Bar = 100 µm

63.6 *Evadne***, Gravid Female (Cladocera)**
Scale Bar = 100 µm

Plate 63. Crustacea: Subclass Ostracoda (Ostracods) and Order Cladocera (Water fleas)

appendages are easily covered by the bivalve carapace, though they extend from the carapace when the animal is crawling. Some common members of coastal plankton in the North Pacific are the genus *Conchoecia* (Figure 62.1) and representatives of the family Cypridinidae (Figure 62.6). Some members of the family Cypridinidae are known to bioluminesce. See Plates 62 & 63 for illustrations of various marine ostracods.

Order Cladocera (Water Fleas)

This group of crustaceans is best known by the freshwater *Daphnia*, which is often used in laboratories as food for creatures such as the cnidarian *Hydra*. Members of the order Cladocera (class Branchiopoda) characteristically exhibit a large, anterior, single compound eye and an enlarged carapace. The cladoceran carapace covers most of the animal, except for the jointed appendages. Several species of marine *Daphnia*, similar to their freshwater congeners, are regularly found in plankton samples. Common in samples from Central California is this *Daphnia* shown with an elongated posterior spine (Figure 63.3). Other marine cladocerans include *Evadne* (Figure 63.5) and the more Northern Pacific *Podon* (Figure 63.4). Some cladocerans are sexually dimorphic (e.g., *Evadne*, with both male and female illustrated in Plate 63). Females of the genus *Evadne* carry fertilized eggs within a carapace that functions as a brood chamber. In Figure 63.6 the compound eyes of brooded young are visible. Both the female and the male, more conical in appearance than the female, may occur in dense swarms. They usually swim near the ocean's surface and, therefore, may be missed if samples are taken from mid-water.

Subclass Copepoda

The most abundant group of crustaceans, and plankton in general, is the Copepoda (class Maxillopoda). Copepods are often primary herbivores of the abundant photosynthesizers (e.g., diatoms) in the plankton. They possess a single, small compound eye, located centrally in the anterior region (a freshwater genus, *Cyclops*, is aptly named for this characteristic). Most copepods have two jointed antennae that extend laterally from the head, short and stubby in some and long and graceful-looking in others. Copepods have no abdominal appendages. The body shape is generally elliptical and may be laterally compressed, though dorso-ventral compression is common in parasitic forms. Some parasitic forms superficially resemble the Isopoda more than other copepods (Plate 71).

There are ten orders of copepods. The three most common free-living orders are the Calanoida, the Harpacticoida and the Cyclopoida. One common order of parasitic copepods are the Siphonostomatoida (Plate 70). Other parasitic copepods are illustrated in Plate 71. Most orders are distinguishable by characteristic body shapes.

Calanoid copepods, usually the most abundant type of copepod far away from the shore, have an ovoid body with a moveable joint behind the sixth and last segments of the thorax (Plates 66 & 67). Antennae tend to be long and the abdomen generally has four segments, located prior to the telson. Eggs, when present on the females, are carried in a single cluster.

Harpacticoid copepods (Plate 69), more abundant in demersal, benthic and near shore samples, have a less noticeable division between body regions, do not reach as large a size as some calanoid copepods, and have shortened antennae. Shorter antennae provide less surface area for flotation, but allow more streamlined navigation of tight spaces.

Cyclopoid copepods, represented by only a few genera (e.g., *Oithona*, Plate 68), have shortened antennae and a movable joint between the fifth and sixth segments. Eggs, when present, are stored in paired sacs. The abdomen, or **urosome**, has five or six segments, sometimes fused, anterior to the telson.

There are several stages of nauplius larva displayed in the Copepoda, including nauplius I, nauplius II and the **metanauplius**. Examples of copepod developmental stages are given in Plate 65. Newly metamorphosed calanoid juveniles are illustrated in Figures 66.1 & 66.2 (note the reduced abdomen when compared to mature calanoids). Calanoid taxonomy is based upon the number of segments behind the head region. Those calanoids illustrated in Plate 66 have three segments behind the head, while most of those shown in Plate 67 have four or five. Examples of cyclopoid and harpacticoid copepods are given in Plates 68 & 69. One harpacticoid (*Tigriopus*, Figure 69.4) is found in intertidal pools in the upper splash zone. These pools receive input of fresh seawater only when waves break at high tide, splashing ocean water into rock pools just above the littoral zone. *Tigriopus* must have a wide range of physiological tolerances to be able to withstand the variety of temperatures and salinities that splash pool organisms are subjected to.

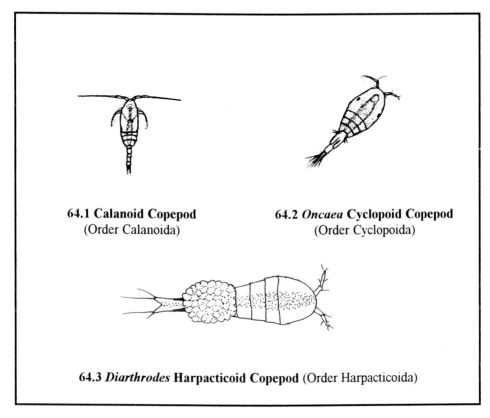

64.1 Calanoid Copepod
(Order Calanoida)

64.2 *Oncaea* Cyclopoid Copepod
(Order Cyclopoida)

64.3 *Diarthrodes* Harpacticoid Copepod (Order Harpacticoida)

Plate 64. Representatives of the three common orders of pelagic marine Copepoda

A useful guide to identifying calanoid and cyclopoid copepods is Fulton (1968). Lang (1965) is one of the best references for detailed identification of harpacticoid copepods. Many parasitic forms are described in Humes & Stock (1973). Gardner & Szabo (1982) provides an identification of marine pelagic copepods from the region of British Columbia.

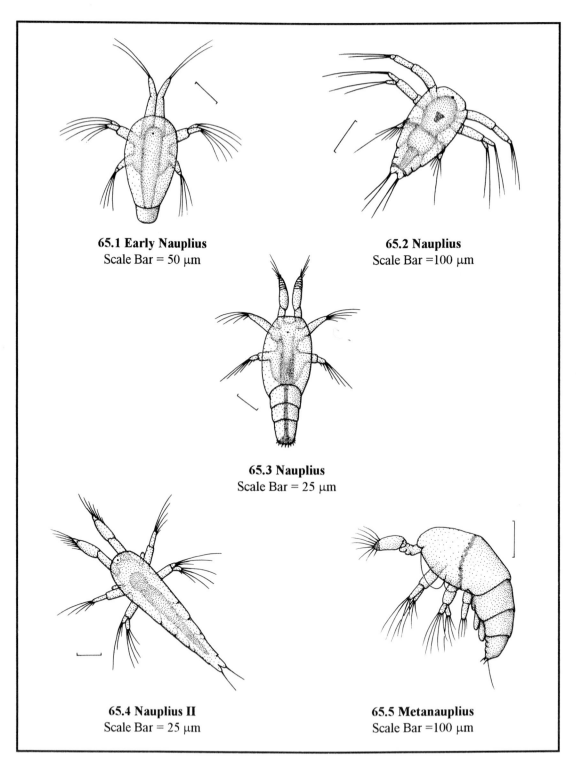

Plate 65. Crustacea: larval stages in the development of some Copepoda

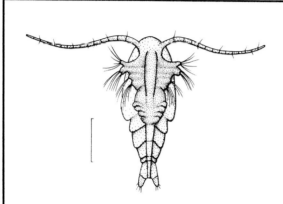

66.1 Calanoid Juvenile, Ventral View
Scale Bar = 250 μm

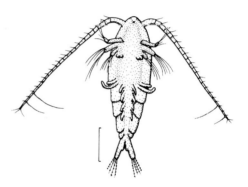

66.2 Calanoid Juvenile, Ventral View
Scale Bar = 250 μm

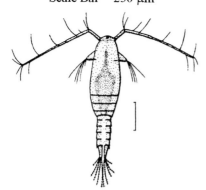

66.3 *Acartia* **(Calanoid)**
Scale Bar = 250 μm

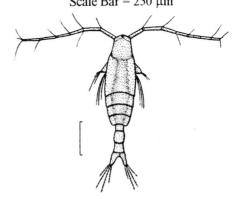

66.4 *Acartia* **(Calanoid)**
Scale Bar = 250 μm

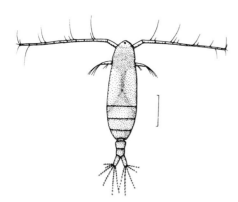

66.5 Unidentified Calanoid
Scale Bar = 250 μm

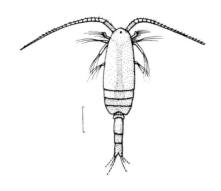

66.6 *Paracalanus* **(Calanoid)**
Scale Bar = 250 μm

Plate 66. Crustacea: Calanoid Copepods (Order Calanoida)

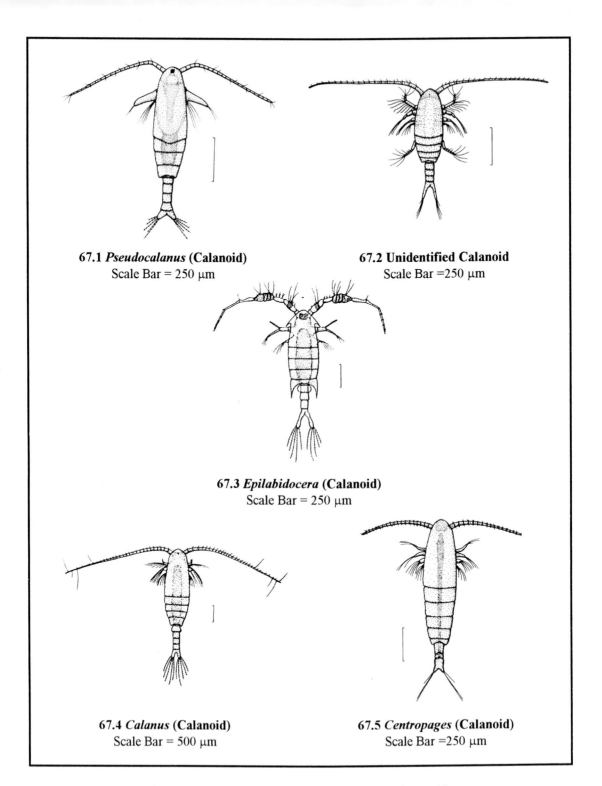

Plate 67. Crustacea: Calanoid Copepods (Order Calanoida)

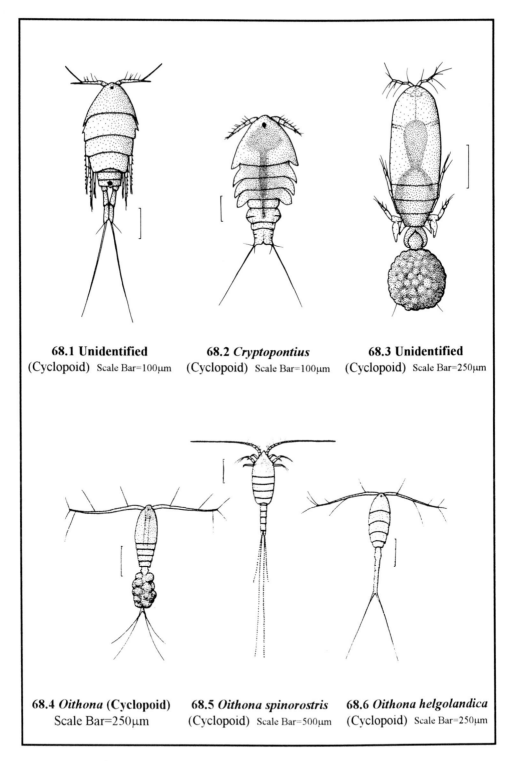

Plate 68. Crustacea: Cyclopoid Copepods (Order Cyclopoida)

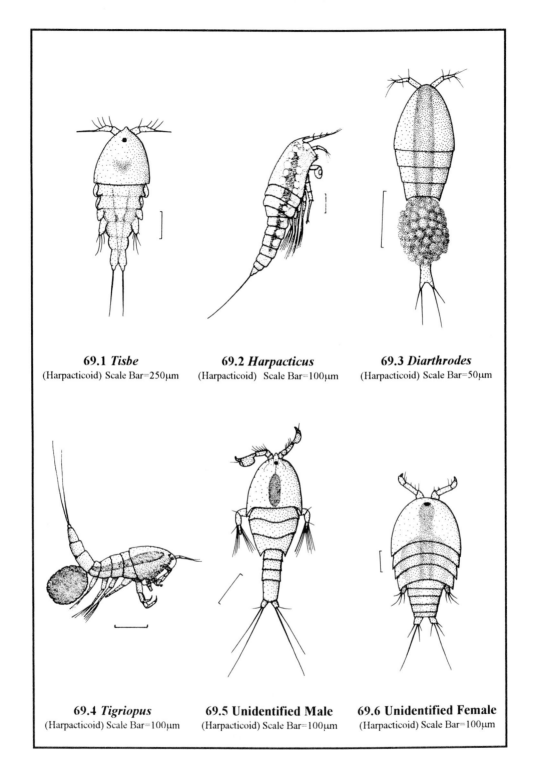

Plate 69. Crustacea: Harpacticoid Copepods (Order Harpacticoida)

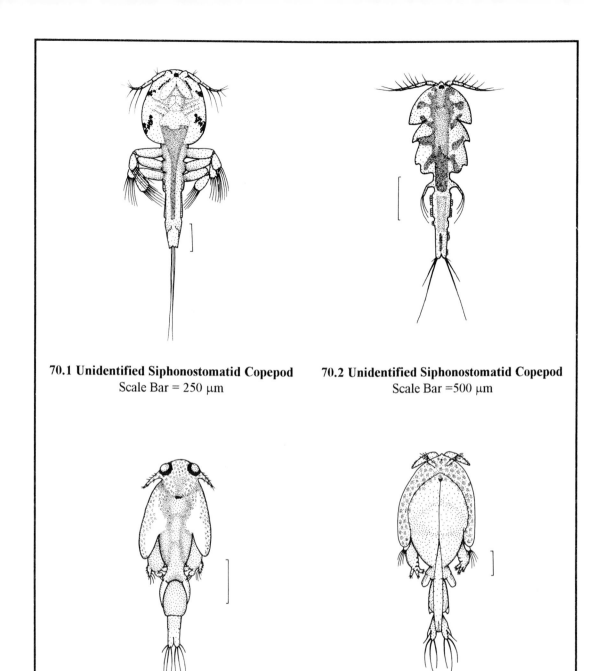

70.1 Unidentified Siphonostomatid Copepod
Scale Bar = 250 μm

70.2 Unidentified Siphonostomatid Copepod
Scale Bar = 500 μm

70.3 *Caligus*
Scale Bar = 1 mm

70.4 Unidentified Siphonostomatid Copepod
Scale Bar = 500 μm

Plate 70. Crustacea: Parasitic Copepods, Order Siphonostomatoida

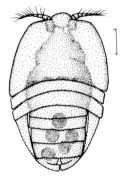

71.1 *Peltidium*, Female (Harpacticoid)
Scale Bar = 100 μm

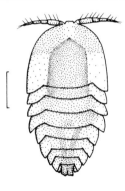

71.2 *Sapphirina*, Male (Cyclopoid)
Scale Bar = 250 μm

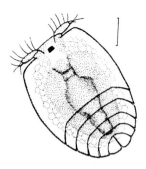

71.3 *Porcellidium* (Harpacticoid)
Scale Bar = 250 μm

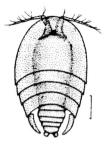

71.4 Unidentified Parasitic Copepod
(Harpacticoid)
Scale Bar = 250 μm

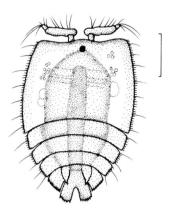

71.5 Unidentified Parasitic Copepod
Scale Bar = 500 μm

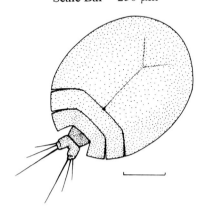

71.6 Unidentified Parasitic Copepod
Scale Bar = 250 μm

Plate 71. Crustacea: Parasitic Copepods

Class Malacostraca (Crabs, Shrimps and their relatives)

Most of balance of planktonic crustaceans are assigned by zoologists to the class Malacostraca. Within this group are two important superorders with many planktonic representatives: the Peracarida and the Eucarida. These two superorders include the true crabs, the various shrimp groups, lobsters, amphipods, isopods and cumaceans. Most members of the class Malacostraca have twenty **somites**: five in the head region, eight in the thorax and seven in the abdominal section. A carapace is present in all but the Isopoda and Amphipoda. Malacostracans have around nineteen paired appendages. Abdominal appendages are **pleopods** and are used in respiration, locomotion, or for the attachment of eggs. Thoracic appendages are the **periopods**. When a heavy claw is present, it is known as a **chela**. We divide this group into manageable sections where general appearance and life histories are discussed.

Superorder Peracarida (Class Malacostraca)

Ten orders of crustacea are found within this superorder. Those commonly represented in the plankton include the orders Mysida, Cumacea, Isopoda, Amphipoda and Tanaidacea. In the adult representatives of these orders, some of which are planktonic and some of which are benthic, the carapace, when present, does not entirely cover the thorax and young are often brooded ventral to the thorax.

Order Mysida (Mysid Shrimps). These shrimp-like forms have elongate bodies, a thin carapace, stalked compound eyes, eight biramous thoracic appendages and a thoracic brood pouch. They often have statocysts located in the **endopods.** They are rapid swimmers, using their fast-moving thoracic appendages, and are marvelous organisms to study when found in the plankton. They are often transparent and the contraction of the tubular heart, attendant blood circulation, movement of mouth parts and peristaltic action of the digestive system can be observed through the body wall. Some planktonic mysid shrimp genera are illustrated in Plates 72 & 73. A detailed key to the mysid shrimps of the Northeastern Pacific is provided by Banner (1950).

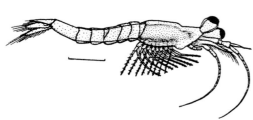

72.1 *Neomysis*, Lateral View
Scale Bar = 500 μm

72.2 *Stilomysis*, Lateral View
Scale Bar = 250 μm

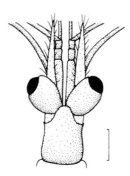

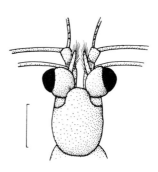

72.3 *Neomysis*, Dorsal Anterior View
Scale Bar = 250 μm

72.4 *Stilomysis*, Dorsal Anterior View
Scale Bar = 250 μm

Plate 72. Crustacea: Mysid Shrimps (Order Mysida)

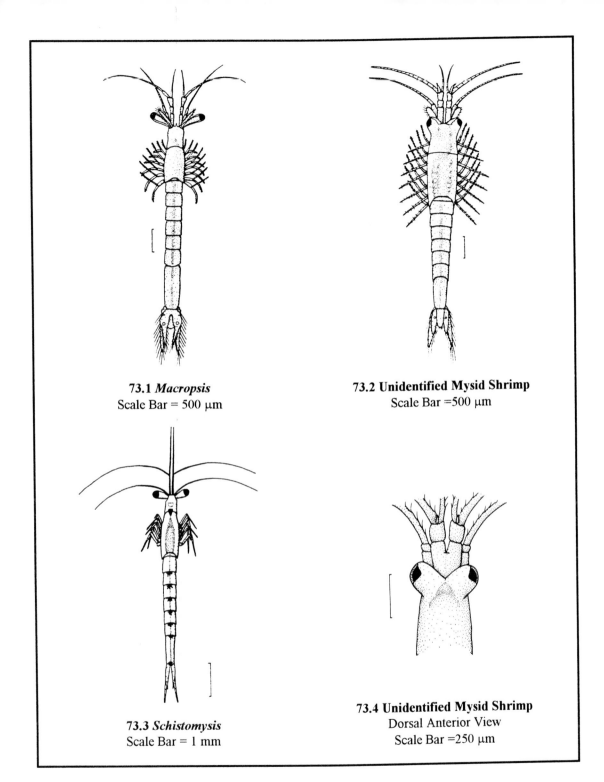

73.1 *Macropsis*
Scale Bar = 500 μm

73.2 Unidentified Mysid Shrimp
Scale Bar = 500 μm

73.3 *Schistomysis*
Scale Bar = 1 mm

73.4 Unidentified Mysid Shrimp
Dorsal Anterior View
Scale Bar = 250 μm

Plate 73. Crustacea: Mysid Shrimps (Order Mysida)

Order Cumacea. Cumaceans are primarily bottom-dwelling organisms that feed on suspended detritus or benthic organic films. Their long, slender abdomen often lacks appendages. This narrow abdomen, emerging from a bulbous cephalothoracic region, combined with the **styliform** telson and uropods, gives the cumacean a very unique appearance. Eyes are close-set on either side of the **rostrum**. Periopods are not biramous in cumaceans. Some cumaceans from Central California are illustrated in Figures 74.1 & 74.2.

Order Tanaidacea. The tanaids are a small group of seldom seen crustaceans, though at times they may be taken in plankton tows. The body of tanaids is elongate and cylindrical with paired appendages throughout the length of the body and large chelae-like **gnathopods** (Figure 74.3). See Hatch (1947) for descriptions of tanaids from the Puget Sound and the coast of Washington State.

Order Isopoda. This order shows no evidence of a carapace. Isopods are generally flattened dorso-ventrally. Perhaps the best known isopods are the terrestrial "pill bugs". Marine isopods in some instances look very similar to these, though not as heavily rounded or dome-shaped. Isopod eyes are unstalked and gills are located on the abdominal appendages. Eggs, which are carried in the ventral thoracic brood pouch, have direct development leading to the release of young very similar to the adult form. Therefore, it is the adult isopod that will occasionally be found in the plankton. Often these are benthic forms that have been swept into the plankton. They are commonly found with floating debris in coastal waters. Some are parasitic on fishes and crabs (e.g., *Bopyrus*) producing enlarged, swollen tumors under the carapace. Some of the isopods occasionally found in Northeastern Pacific plankton include *Gnorisphaeroma*, *Idotea* and *Munna*. These and other isopods are illustrated in Plate 75. Figure 75.1 is what is known as the **cryptoniscid** larva of a parasitic isopod. In some parasitic forms, the egg develops into a **microniscid**, then a cryptoniscid, and finally metamorphoses into the adult. In parasitic forms appendages are often lost, especially in the larval stages. Helpful references on the systematics of isopods include Richardson (1905), Schultz (1969) and Hatch (1947).

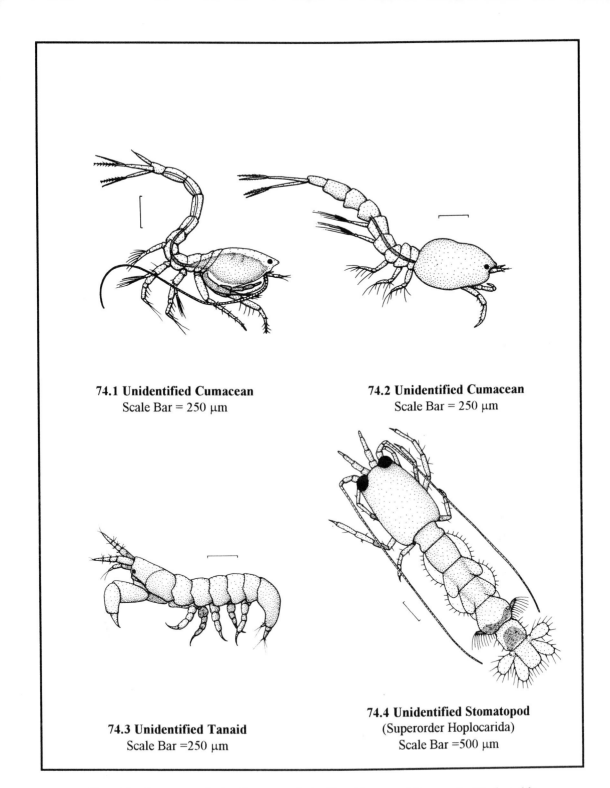

Plate 74. Crustacea: Order Cumacea, Order Tanaidacea and Superorder Hoplocarida

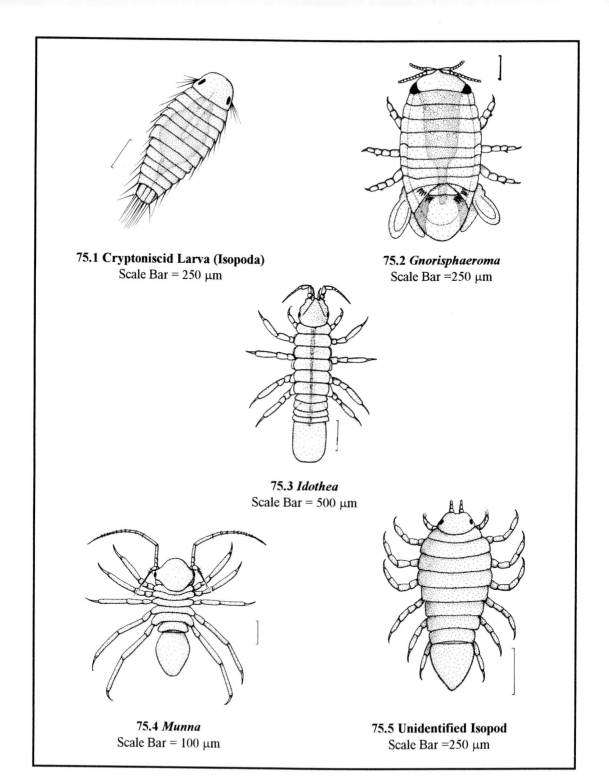

75.1 Cryptoniscid Larva (Isopoda)
Scale Bar = 250 μm

75.2 *Gnorisphaeroma*
Scale Bar = 250 μm

75.3 *Idothea*
Scale Bar = 500 μm

75.4 *Munna*
Scale Bar = 100 μm

75.5 Unidentified Isopod
Scale Bar = 250 μm

Plate 75. Crustacea: Order Isopoda

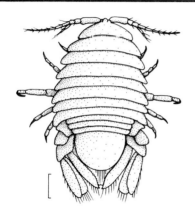

76.1 Unidentified Isopod
Scale Bar = 250 μm

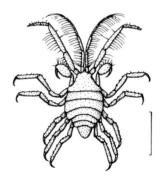

76.2 Unidentified Isopod
Scale Bar = 1 mm

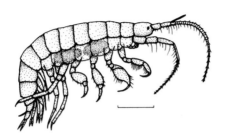

76.3 Unidentified Gammarid Amphipod
Scale Bar = 500 μm

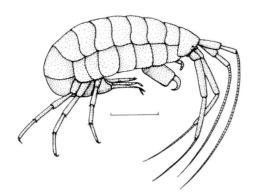

76.4 Unidentified Gammarid Amphipod
Scale Bar = 100 μm

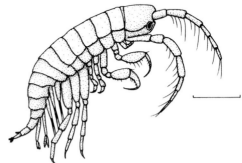

76.5 Unidentified Gammarid Amphipod
Scale Bar = 500 μm

Plate 76. Crustacea: Order Isopoda and Order Amphipoda

Order Amphipoda. In many ways similar to the isopods, these animals are laterally compressed rather than dorso-ventrally flattened. Amphipods also often have their abdominal region slightly flexed or tucked, giving the body a sickle shape. Well-known examples of this order are the beach-hoppers, sometimes called "sand fleas", which burrow in the sand or live in decaying matter on the beach. They often feed upon decaying animal, algal and/or plant matter.

Examples of planktonic amphipods include *Phronima* and *Hyperia* (suborder Hyperiidea) which are often associated with larger jelly plankton (e.g., *Hyperia* has been found living on the pelagic scyphozoans *Aurelia* and *Cyanea* and *Phronima* is found in the tunics of deep water salps). Some may even be found on ctenophores, as was the case with the individual illustrated in Figure 77.6. Hyperiids are recognizable by their larger heads and eyes (Figures 77.5 & 77.6). See Bowman & Gruber (1973) for assistance with the systematics of this suborder.

Gammarid amphipods (suborder Gammaridea) are similar to the hyperiids, but with smaller, more proportional eyes. These amphipods are rapid swimmers, taking occasional rests and, in an observation bowl, settling on their sides where the student can observe their characteristic sickle shape. Several genera of gammarid amphipods, the amphipods most often encountered in plankton samples, are illustrated in Plates 76 & 77. Stebbing (1906) and Barnard (1969) are good sources for detailed information on the identification of gammarid amphipods.

Caprellid amphipods (suborder Caprellidea) are often called "skeleton shrimp" because of their unique form which is adapted for clinging to filamentous algae, hydroids, etc. They have an elongate body and highly modified abdomen. They are not typically planktonic, but are still often found on flotsam and branching benthic organisms which have broken free and been swept into the plankton. On these drifting substrata, caprellids can be seen holding on with their posterior appendages and nodding sometimes gracefully, sometimes erratically, as they sway back and forth. Gills can be seen on the thoracic appendages and the brood pouch is also ventrally thoracic. With close observation, it may be possible to see the flow of water over the gills. The appendages are sharply hooked to provide a sure hold on the caprellid's resting place (Plate 78). The body

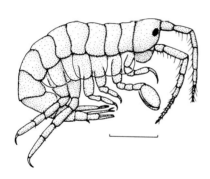

77.1 *Orchestoidea corniculata*
(Gammaridea)
Scale Bar = 500 μm

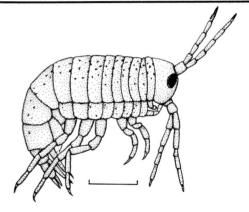

77.2 *Ampithoe* (Gammaridea)
Scale Bar = 500 μm

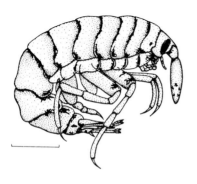

77.3 *Vibilia* (Gammaridea)
Scale Bar = 500 μm

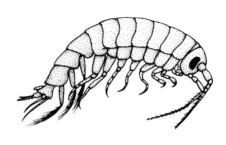

77.4 *Polycheria* (Gammaridea)

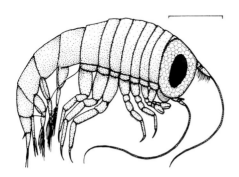

77.5 *Hyperoche* (Hyperiidea)
Scale Bar = 500 μm

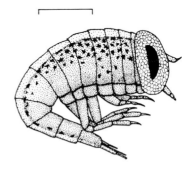

77.6 Hyperiid Amphipod found on Ctenophore
Scale Bar = 500 μm

Plate 77. Crustacea: Order Amphipoda

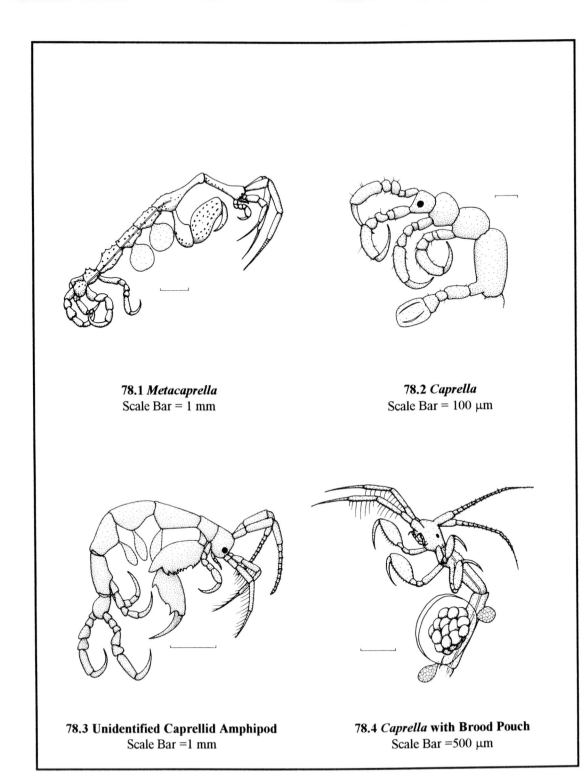

78.1 *Metacaprella*
Scale Bar = 1 mm

78.2 *Caprella*
Scale Bar = 100 μm

78.3 Unidentified Caprellid Amphipod
Scale Bar =1 mm

78.4 *Caprella* with Brood Pouch
Scale Bar =500 μm

Plate 78. Crustacea: Caprellid Amphipods ("Skeleton Shrimp")

shape and the way the caprellids hold their anterior appendages is reminiscent of the familiar praying mantis insect. See Laubitz (1970) for more detailed information on the caprellid amphipods of the North Pacific region.

Superorder Hoplocarida (Class Malacostraca)

Mantis shrimps (order Stomatopoda) have some general characteristics very similar to the specimen in Figure 74.4. Though unidentified, this specimen taken in a plankton tow in Central California does bear the broad, flat carapace, elongated abdomen and gill-bearing appendages found in the stomatopods. In certain regions of the world stomatopods are quite common and their larvae, in some cases morphologically similar to the adults, can regularly be found in the plankton.

Superorder Eucarida (Class Malacostraca)

In this group, the carapace is large and covers the entire thorax. Eyes are set on stalks, gills are thoracic and eggs are carried underneath the abdomen on swimmerets. Development is complex and often involves several larval stages. Most eucarid species belong to the order Decapoda, while just under one hundred described species belong to the order Euphausiacea.

Order Euphausiacea (Superorder Eucarida). Adult euphausids are not usually taken in samples from Central California, though the larval forms do appear occasionally. Euphausids are the well-known krill which are sometimes seen in swarms, especially in the more polar, colder, off-shore waters. Several different euphausid larva stages are illustrated in Plate 79. The genera most likely to be found in the Northeastern Pacific include *Euphausia* and *Thysanoessa* (Banner, 1950).

Order Decapoda (Superorder Eucarida). This order includes the lobsters, crabs and shrimps. The carapace covers the entire thorax and the thoracic appendages are termed periopods, or walking legs, with the first periopod pair bearing chelae. Embryos are often held by abdominal appendages, the pleopods. Lobsters and shrimps both have an extended abdomen with a tail-fan made up of the telson and uropods. Crabs possess

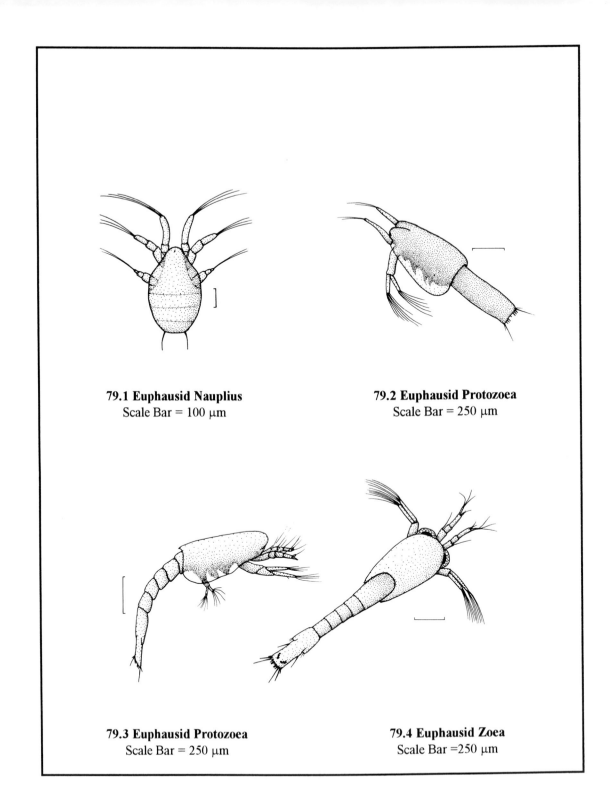

Plate 79. Crustacea: Developmental Stages of Euphausid Shrimp (Order Euphausiacea)

a shortened, modified abdomen folded back underneath the cephalothorax. Gills are contained within the gill chambers under the carapace. This order is represented regularly in the plankton by eight groups within the suborders Pleocyemata and Dendrobranchiata (penaeid shrimps): infraorder Anomura (hermit crabs, sand crabs, etc.), infraorder Brachyura ("true" crabs), infraorder Caridea (caridean shrimps), infraorder Thalassinidea (mud shrimps and ghost shrimps), infraorder Palinura (lobsters), infraorder Astacidea (chelate lobsters) and the families Penaeidae and Sergestidae. The ghost and mud shrimps, lobsters and related groups are sometimes collectively classified as section Macrura. A helpful overview of planktonic developmental forms in the order Decapoda (for the region of British Columbia, though within-family developmental patterns are similar throughout the world) is given by Hart (1971).

These infraorders are primarily represented in the plankton by their various larval forms, though adult forms of some caridean and penaeid shrimps may be captured in plankton nets. The shrimps and related groups generally have an elongated, cylindrical abdomen and a tail-fan composed of the telson and uropods. Dorsal views of some of the adults of caridean shrimps represented in the coastal plankton of Central California by planktonic larvae are illustrated in Plates 80 & 81. Plate 82 shows lateral views of some other carideans. The compound eyes are generally stalked and the eye stalks and carapace often bear **chromatophores** with pigments of red or black. The rostrum can be elongated or greatly reduced. Penaeid shrimps tend to be found at lower latitudes and have three pairs of chelate appendages, while the caridean shrimps lack chelae on the third appendage. Some descriptions of planktonic larvae from representative shrimp species are helpful to get an idea of what developmental forms are like in some of the shrimp groups. Larval stages in several genera of caridean shrimps (family Hippolytidae) are described in Pike & Williamson (1960) and Haynes (1981). Larval development in three families of Caridea is illustrated in Haynes (1985). Early developmental stages in the penaeid shrimp genus *Penaeus* are described in Dobkin (1961) and Cook and Murphy (1971).

There are no lobsters in the waters of the Northeastern Pacific. However, in regions that have lobsters (e.g., *Homarus* on the East Coast

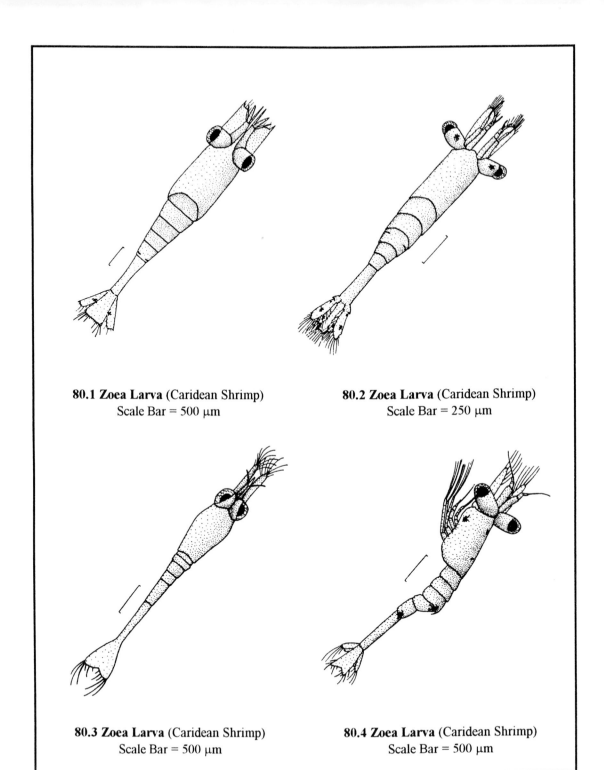

80.1 Zoea Larva (Caridean Shrimp)
Scale Bar = 500 μm

80.2 Zoea Larva (Caridean Shrimp)
Scale Bar = 250 μm

80.3 Zoea Larva (Caridean Shrimp)
Scale Bar = 500 μm

80.4 Zoea Larva (Caridean Shrimp)
Scale Bar = 500 μm

Plate 80. Crustacea: Caridean Shrimp Larvae (Order Decapoda, Infraorder Caridea)

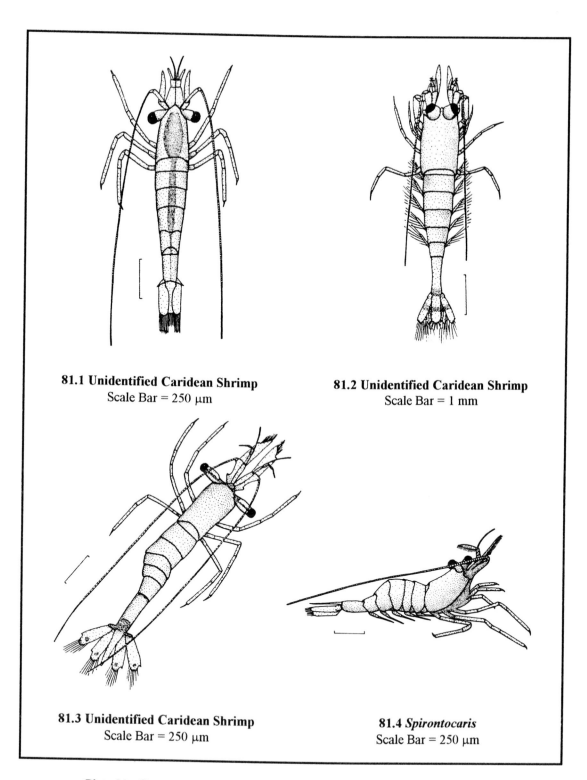

81.1 Unidentified Caridean Shrimp
Scale Bar = 250 µm

81.2 Unidentified Caridean Shrimp
Scale Bar = 1 mm

81.3 Unidentified Caridean Shrimp
Scale Bar = 250 µm

81.4 *Spirontocaris*
Scale Bar = 250 µm

Plate 81. Crustacea: Caridean Shrimps (Order Decapoda, Infraorder Caridea)

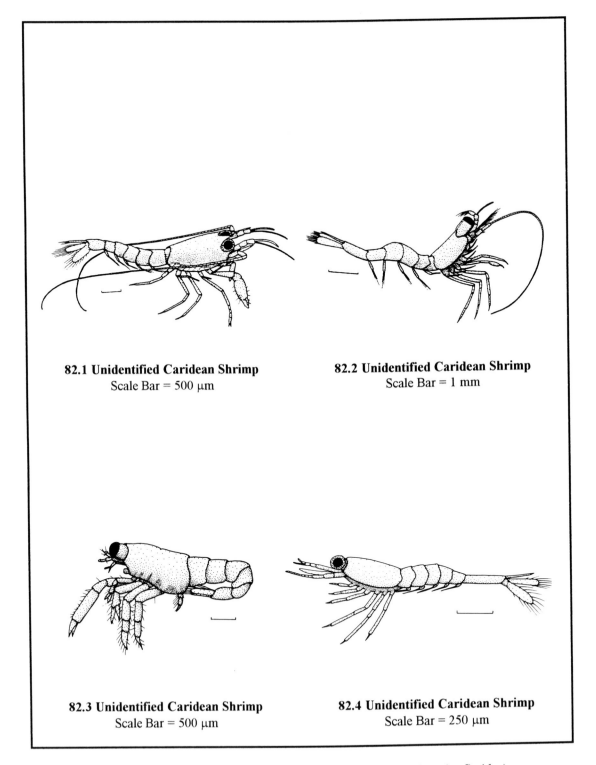

Plate 82. Crustacea: Caridean Shrimps (Order Decapoda, Infraorder Caridea)

and *Panulirus* in Southern California and Mexico) their larvae, including zoea and phyllosoma stages, are regularly found in plankton samples.

Infraorder Anomura (Order Decapoda). Anomurans include the hermit crabs (family Paguridae), the sand crab *Emerita* and the porcelain crabs (family Porcellanidae). Development varies slightly from group to group. In the pagurids (Plates 83 & 84), a **zoea** (rather shrimp-looking) follows the prezoea, called in some groups the **protozoea**. The **megalopa** (Plate 87) is the final planktonic stage before settlement and metamorphosis. The zoea of the sand crab *Emerita* lacks all spines, with the exception of the rostral spine, on the helmet-looking carapace (Figure 85.3). Another sand crab, *Blepharipoda* bears a unique planktonic zoea with a long, slender abdomen emerging from beneath the carapace and a long rostral spine (Figure 85.1). Porcelain crab zoeae (Figure 85.2) are distinct in that they bear dramatically long and straight spines extending from the back of the carapace and the rostrum. A key to planktonic larval stages of British Colombia Anomura is given by Hart (1937). Early zoeal stages of some lithodid anomurans (Family Lithodidae) are described by Haynes (1984).

Infraorder Brachyura (Order Decapoda). Brachyuran, or "true" crabs include the common genus *Cancer*, which encompasses the familiar rock crabs and many of the commercially harvested species (e.g., *Cancer magister*, the dungeness crab). Kelp crabs, shore crabs and decorator crabs are also classified among the Brachyura. Brachyurans release an early **zoea** from eggs carried on the pleopods of the female. These free-swimming larvae soon molt to the next zoeal stage with the typical zoeal helmet-like carapace with spines and an elongate rostrum (Plate 86). The brachyuran settling stage is the megalopa (e.g., Plate 88). The megalopa is much closer in morphology to the adult crab than the zoea. The abdomen, while having the ability to be flexed under the carapace, can still extend behind the megalopa where the abdominal appendages make this larval stage an especially effective swimmer. If larvae are maintained in the laboratory, the researcher will be able to observe the metamorphosis of a megalopa larva to a juvenile crab. Abundant literature exists on the unique characteristics of planktonic brachyuran larvae within a species. We attempt here to mention the more comprehensive works as well as some

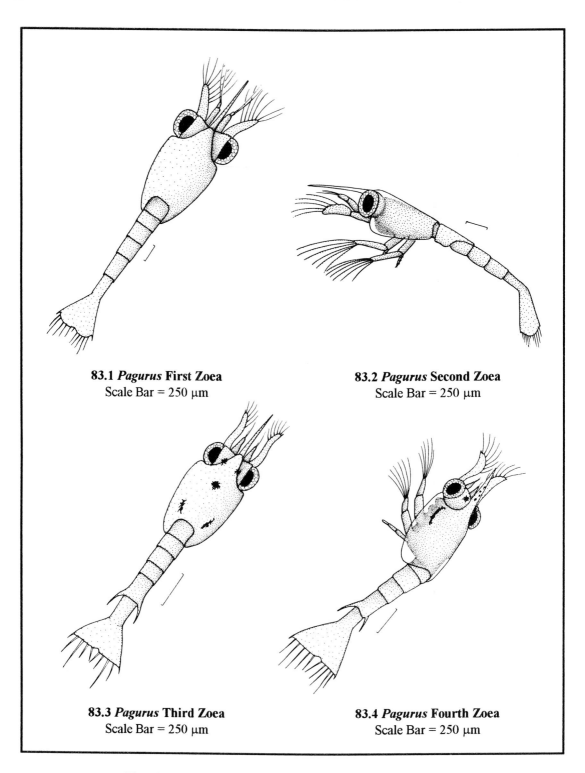

83.1 *Pagurus* First Zoea
Scale Bar = 250 μm

83.2 *Pagurus* Second Zoea
Scale Bar = 250 μm

83.3 *Pagurus* Third Zoea
Scale Bar = 250 μm

83.4 *Pagurus* Fourth Zoea
Scale Bar = 250 μm

Plate 83. Crustacea: Larval Stages of the Hermit Crab *Pagurus*

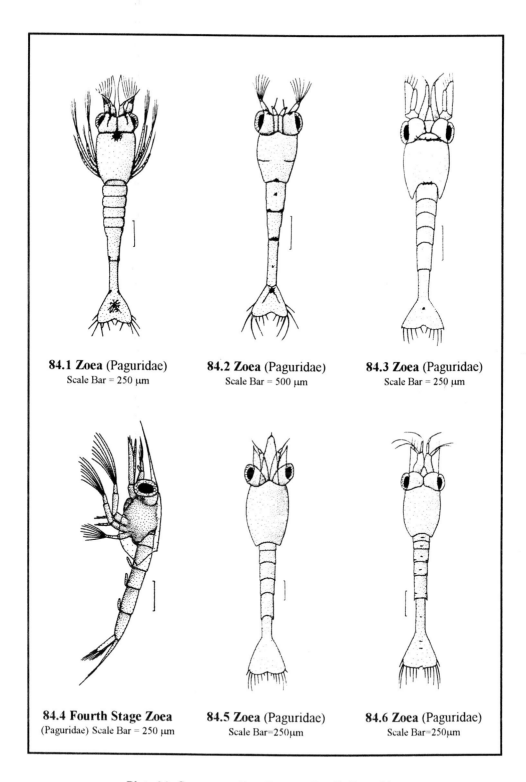

Plate 84. Crustacea: Zoea Larvae, Family Paguridae

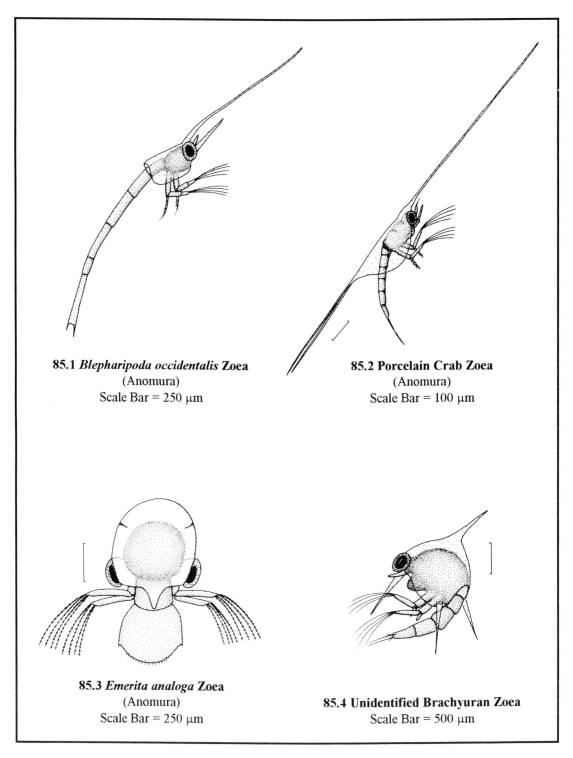

85.1 *Blepharipoda occidentalis* Zoea
(Anomura)
Scale Bar = 250 μm

85.2 Porcelain Crab Zoea
(Anomura)
Scale Bar = 100 μm

85.3 *Emerita analoga* Zoea
(Anomura)
Scale Bar = 250 μm

85.4 Unidentified Brachyuran Zoea
Scale Bar = 500 μm

Plate 85. Crustacea: Zoea Larvae of Anomuran and Brachyuran Crabs (Order Decapoda)

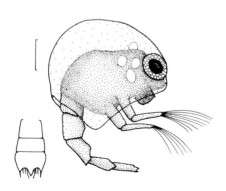

86.1 Unidentified Pre-Zoea
Scale Bar = 250 μm

86.2 Unidentified Pre-Zoea
Scale Bar = 250 μm

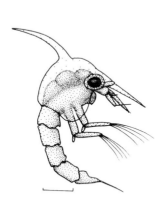
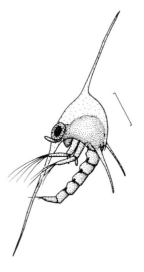

86.3 Unidentified Zoea
Scale Bar = 500 μm

86.3 Unidentified Zoea
Scale Bar = 500 μm

Plate 86. Crustacea: Larval Stages of Brachyura (Order Decapoda)

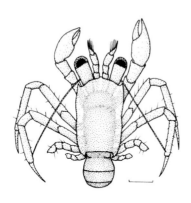

87.1 Pagurid Megalopa (Anomura)
Scale Bar = 1 mm

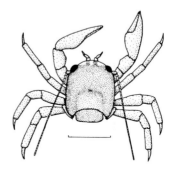

87.2 Porcellanid Megalopa (Anomura)
Scale Bar = 1 mm

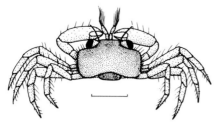

87.3 Unidentified Brachyuran Megalopa
Scale Bar = 1 mm

87.4 Unidentified Brachyuran Megalopa
Scale Bar = 1 mm

Plate 87. Crustacea: Megalopae of Anomuran and Brachyuran Crabs (Order Decapoda)

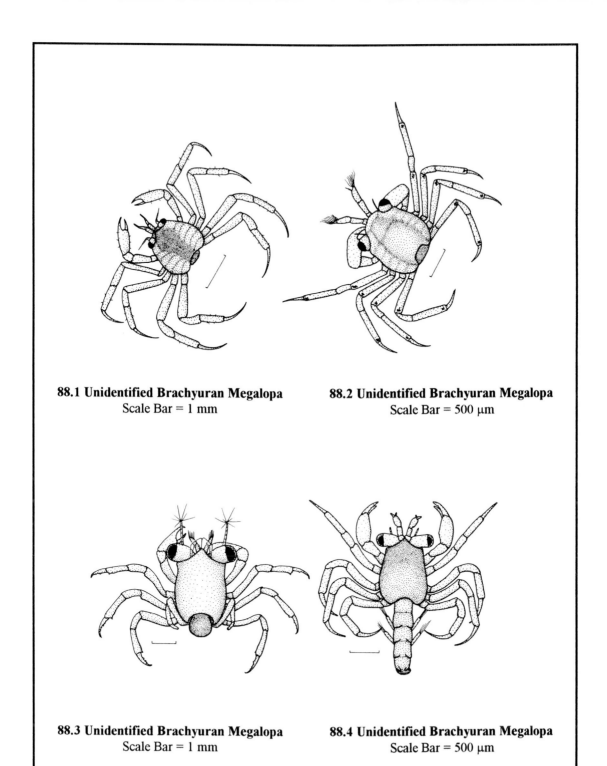

Plate 88. Crustacea: Megalopae of Brachyuran Crabs (Order Decapoda)

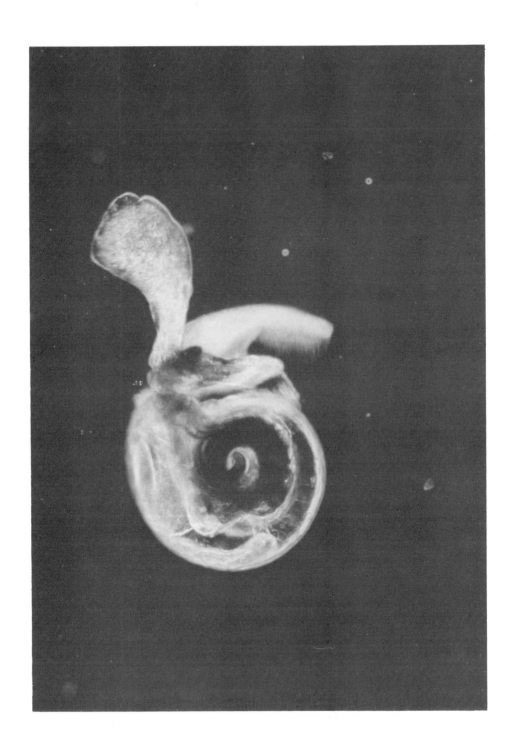

references on genera of common interest. Rice (1980) analyzes the morphology of brachyuran zoeae. Hart (1935, 1960) provide descriptions of larval development in representatives of four brachyuran families. Hyman (1926) presents studies on the larvae of xanthid crabs (family Xanthidae), while Lebour (1927, 1928) describes larvae from three different genera of brachyura. Other works with descriptions of planktonic larvae include the following: Knudsen (1958, 1959, 1960), Poole (1966) for *Cancer magister*, Trask (1970) for *Cancer productus*, Diaz & Costlow (1972), Ally (1974, 1975), Roesijadi (1976) for three *Cancer* species, Schlotterbeck (1976) for *Pachygrapsus crassipes*, Johns & Lang (1977) for *Libinia*, Bousquette (1980) for *Pinnixa* and Anger et al. (1990) for *Uca*.

Phylum Mollusca

There are eight **extant** classes of this phylum. Many of these are regularly represented in coastal plankton samples. Planktonic larval forms from the classes Polyplacophora, Gastropoda, Bivalvia and Cephalopoda are often found in coastal plankton samples. There are also adult planktonic molluscs, primarily from the class Gastropoda.

Class Polyplacophora (Chitons)

Most chitons have an unusual egg that stands out in plankton samples due to the encasing ornate **chorion**. This chiton egg (Figure 89.1) collected in Central California by Professor Smith remained an unknown in his plankton files for several years before being identified as an egg from the class Polyplacophora. Fertilized chiton eggs develop into early free-swimming trochophores (Figure 89.2) and then on into a late trochophore. As seen in Figure 89.3, eyespots are sometimes present just posterior to the prototroch. Late chiton trochophores, if maintained and observed in the laboratory for several days, may begin to show early Plate segmentation in preparation for metamorphosis and settlement. Some invertebrate larvae appear to delay settlement in the absence of a suitable adult habitat. One example of this in polyplacophoran larvae is the planktonic trochophore of

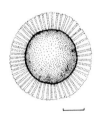

89.1 Chiton Egg
Scale Bar = 100 μm

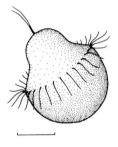

89.2 Early Chiton Trochophore
Scale Bar = 100 μm

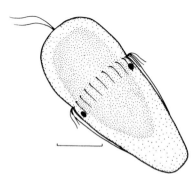

89.3 Chiton Trochophore
Scale Bar = 100 μm

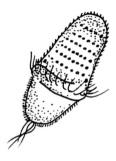

89.4 Late Chiton Trochophore
(Adapted from Watanabe & Cox, 1975)

89.5 Chiton Metamorphosis
(Adapted from Watanabe & Cox, 1975)

89.6 Juvenile Chiton
(Adapted from Watanabe & Cox, 1975)

Plate 89. Mollusca: Chiton Developmental Stages (Class Polyplacophora)

the chiton *Tonicella lineata*. Adult *Tonicella* normally live in rocky shoreline pools and crevices where encrusting corraline algae is present. Planktonic larvae remain at the late trochophore stage unless corraline algae, or corraline algae extract, is present in the water. Corraline algae is a primary food source for adults of this species of chiton. This larval response to the presence or absence of the food of the adult helps ensure that the trochophore does not settle in a place where the adult cannot survive. An overview of development in this class, including illustrations or pictures of trochophores from six genera, is given in Pearse (1979).

Class Gastropoda (Snails and their relatives)

Planktonic forms of the gastropods include pelagic adult forms, larvae of many benthic species and occasionally the egg cases of some snails, limpets, nudibranchs, etc.. A common egg case found in the plankton is that of *Littorina* (Figure 90.2). Littorine egg cases are disk-shaped, flattened and can be as large as 1 mm in diameter. A dozen or more developing embryos, in various stages of cleavage, or later stage veliger larvae, may be found within the case. These veligers eventually hatch out to begin their life in the plankton. Another unidentified egg case is shown in Figure 90.3.

The basic larva of most gastropods is the veliger. The veliger is named for the velum, a broad expanded sheet of mantle tissue. The velum is ciliated at the outer border and is very flexible, allowing it to be retracted within the shell. Some veligers have pigments in the four corners of the velum. Prosobranch veligers often stand out in plankton samples. When these larvae are disturbed, the velum is retracted and they drop quickly to the bottom of the dish or jar. Here they can easily be picked out by looking for their coiled shells among the settled debris. If the container is left undisturbed, veligers may extend their velum from the shell and begin to swim again. Unidentified veligers are illustrated in Plates 90 & 91. The literature identifying prosobranch veligers to genus is a scattered collection of references from seas around the world. These references, though from many different regions, may be helpful to those trying to identify veligers to genus. For those interested in genera found in the tropics, Taylor (1975)

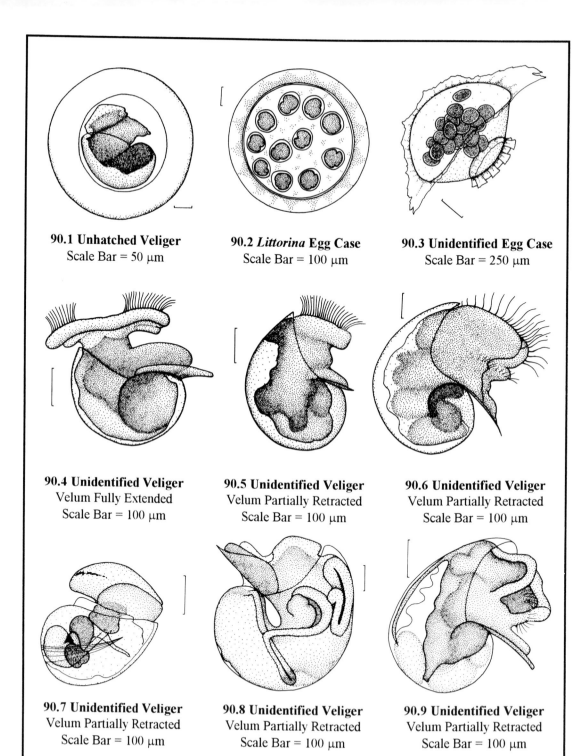

Plate 90. Snail Egg Cases and Veligers (Class Gastropoda)

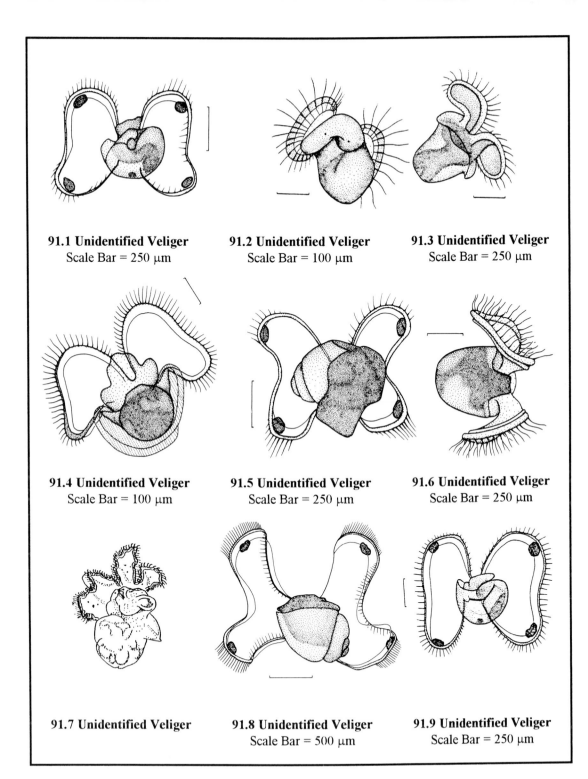

Plate 91. Snail Veligers (Class Gastropoda)

provides some illustrations of veliger shells. The velum is occasionally drawn, but identification using this thesis relies primarily upon shell shapes and proportions. Fretter & Pilkington (1970) provides a key to some prosobranch veligers. These beautiful illustrations make it possible to make identifications using the velum characteristics, as well as the shell shape. An overview of the use of larval characteristics in systematics of marine prosobranchs and the consequent importance of larval descriptions is given by Robertson (1974). Thorson (1940) presents descriptions of developing veligers and eggs masses of both prosobranchs and opisthobranchs in the Iranian Gulf. Some of the genera studied are widespread in distribution. Drawings and photographs from Amio (1963) may also be helpful for identifying the veligers of some gastropods. Excellent pictures of veliger larvae from the Gulf of Naples, including genera with wide distributions, are available in Richter & Thorson (1975). Some good photographs of living veligers with the velum extended are also published in this article. Thiriot-Quiévreux (1980, 1983) and Thiriot-Quiévreux & Scheltema (1982) use close-up photographs of shells to aid in the identification of some prosobranch veligers off the coast of Beaufort, North Carolina and more northerly Atlantic waters.

A slight variation on the veliger is the **echinospira** (Plate 92). Another unique find is *Gastropteron* (order Cephalaspidea), a gastropod with this unique pelagic juvenile form (Figure 93.2). *Gastropteron*, a genus more common in Mediterranean and Atlantic waters, is oceanic and only rarely found in Central California coastal plankton samples. Upon metamorphosis from the veliger into the juvenile, the velum is lost and the mantle is modified to form broad wing-like extensions. The juveniles and adults move by soaring on these undulating wings.

Pterods (sea butterflies), not common in coastal samples, may occasionally be found in abundance when samples are taken in the Summer at the highest tides. Pteropods may be shelled, as in *Limacina* (order Thecosomata, Figure 93.3), or naked, as in *Clione* (order Gymnosomata, Figure 93.4). Both flap their "wings", an action which keeps them suspended in the water column with lurching upward movements.

Members of the subclass Opisthobranchia, including the sea slugs or nudibranchs, also have veliger larvae. Opisthobranch veligers develop in

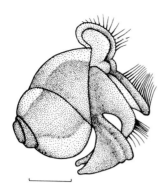

92.1 Unidentified Veliger
Scale Bar = 250 μm

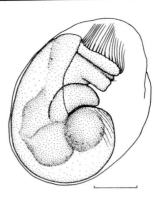

92.2 Echinospiral Veliger Larva
Scale Bar =100 μm

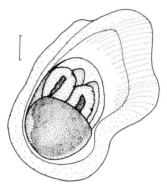

92.3 Echinospiral Veliger Larva
Scale Bar =100 μm

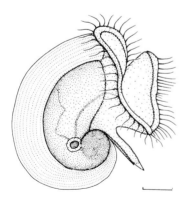

92.4 Echinospiral Veliger Larva
Scale Bar =100 μm

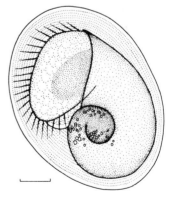

92.5 Echinospiral Veliger Larva
Scale Bar =100 μm

Plate 92. Larval Snails (Class Gastropoda)

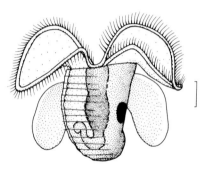

93.1 *Gastropteron* Veliger
Scale Bar = 100 μm

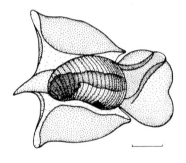

93.2 *Gastropteron* Juvenile
Scale Bar = 250 μm

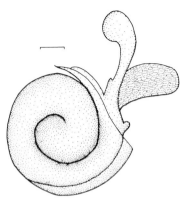

93.3 *Limacina* (Thecosomate Pteropod)
Scale Bar = 100 μm

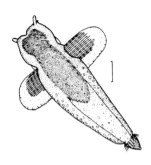

93.4 *Clione* (Gymnosomate Pteropod)
Scale Bar = 100 μm

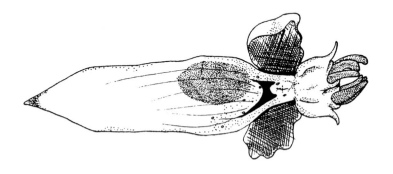

93.5 *Clione* (Gymnosomate Pteropod)
Dorsal View

Plate 93. Pelagic Snails (Class Gastropoda)

egg cases and, in many species, are then released for a time to live a planktonic existence. They tend to be smaller than the larvae of limpets and prosobranch snails and the velum is reduced relative to the shell size when compared to veligers of the other groups. When the shell is present in the veliger it is lost when metamorphosis to the adult form occurs (most adult nudibranchs lack a shell). Several late larval forms and post larvae are shown in Plate 94. The larva and newly metamorphosed juvenile of the green sea hare *Phyllaplysia taylori* are shown in Figures 94.2 & 94.4, respectively. Benthic adult *Phyllaplysia* are found living on the blades of eel grass in tidal mud flats. Studies on the evolutionary significance of opisthobranch veligers are presented in Page (1994). One of the best ways to study opisthobranch development is to obtain egg cases or maintain adults in the laboratory until egg cases are laid. Some literature is available on identifying the species from the egg case characteristics (O'Donoghue & O'Donoghue, 1922; Costello, 1938; Greene, 1968; Kress, 1971). When trying to identify free-swimming opisthobranch veligers from plankton samples, Hurst (1967) is a practical reference, describing twenty-four widely distributed genera. Kress (1972, 1975) include photographs of veligers from several genera of opisthobranchs.

Class Bivalvia (=Pelecypoda)

These molluscs (the clams, oysters, mussels, etc.) are well-known in their benthic adult forms. Like gastropods, bivalves possess a veliger larva. Like the adults, bivalve veligers have two shells, from which the velum opens to provide locomotion for the planktonic larva. Often veligers are transparent and internal anatomy of live larvae can be studied. Larval bivalves possess **adductor muscles**, as the adult bivalves do, which are used to keep the valves closed. These adductor muscles are found internally on either side of the **umbo**, the bump or knoll on the valve. In newly metamorphosed clams, one can see the gills active in respiration as oxygen-laden water flows over them. As the bivalve grows, the pumping heart can be seen through transparent valves. The late veliger grows a typical bivalve foot and becomes a **pediveliger**. Pediveligers can swim up into the water column for transport in ocean currents and then close their

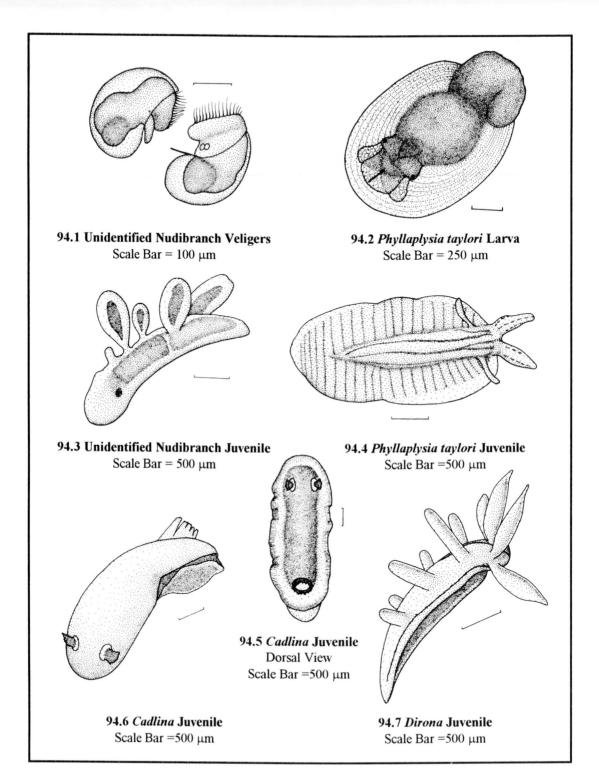

Plate 94. Nudibranch Larvae and Juveniles (Subclass Opisthobranchia)

velum, sinking to the ocean floor. Here the foot tests the sediment and assists in burrowing if the habitat is satisfactory. The bivalve veligers, pediveligers and juveniles pictured in Plate 95 are from the genus *Mytilus*. Many bivalves add visible growth rings as they age. Because of the importance of bivalves to commercial and sport fisheries, some work has compiled photographs and illustrations of these larvae. Some of the most useful guides to larval bivalve identification include Sullivan (1948), Rees (1950), Loosanoff et al. (1966) and Chanley & Andrews (1971). Ockelmann (1962) provides a guide to larval marine bivalves of Europe and Schweinitz & Lutz (1976) use photographs to compare larvae of the genera *Modiolus* and *Mytilus*. Most bivalve identification relies not only upon shell shape and proportion, but upon the size of the veliger or pediveliger. Certain genera may be distinct at a certain size, but resemble other genera of other sizes. In general, veligers of different species tend to morphologically diverge as they grow. Lutz et al. (1982) presents a key to the identification of bivalve larvae using **hinge** structure.

Class Cephalopoda (Octopods and Squids)

Cephalopods lack the typical mollusc larval stages. Often hatching from egg cases, young do live a planktonic life for a time. Many bear a yolk sac to aid in completing early development. Because the young cephalopods closely resemble the adults and lack a dramatic metamorphosis when settling out of the plankton, many think of them as having direct development. As with opisthobranchs, finding egg cases and maintaining them in the laboratory may be the best way to observe cephalopod development. Typically adult cephalopods do not do well in aquaria, though a female octopus will occasionally deposit an egg mass on the underside of a rock or other surface. Egg cases from the squid *Loligo*, sometimes termed "dead man's fingers", are laid by the dozens or hundreds and attached to the sea floor subtidally. Occasionally these egg cases break loose and wash ashore. Maintaining cases in the laboratory is the best way to observe development in squids. Illustrations of *Loligo* in Plate 96 were drawn from specimens obtained in this manner.

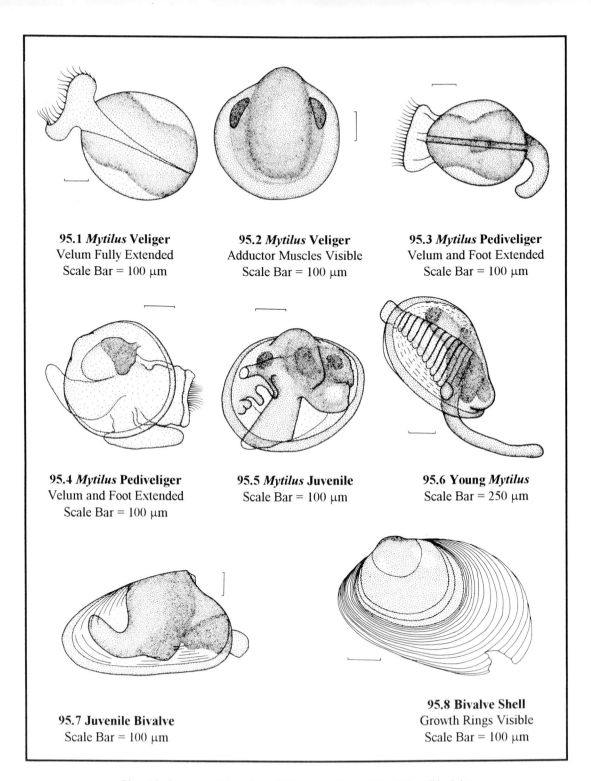

Plate 95. Larvae and Juveniles of Clams and Their Allies (Class Bivalvia)

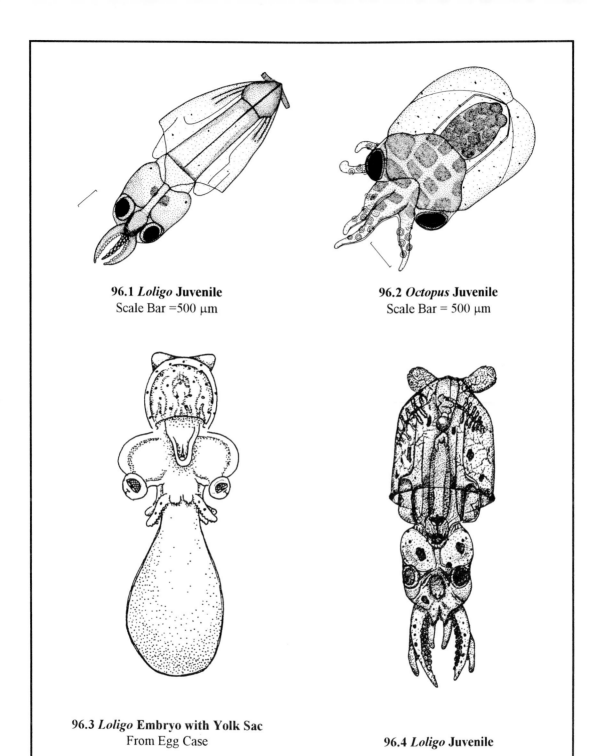

96.1 *Loligo* Juvenile
Scale Bar = 500 μm

96.2 *Octopus* Juvenile
Scale Bar = 500 μm

96.3 *Loligo* Embryo with Yolk Sac
From Egg Case

96.4 *Loligo* Juvenile

Plate 96. Juvenile Squid and Octopus (Class Cephalopoda)

Young squid and *Octopus* (Figure 96.2) periodically turn up in plankton samples. The illustrated *Octopus* specimen was taken in Tomales Bay, California. Most cephalopods, including planktonic juveniles, have chromatophores covering most of the epidermis. These pigment pores can open and close, altering the overall color of the animal. Viewed under a microscope, the bursting and shrinking colors are interesting to observe. Boletzky (1974) reviews the development of cephalopods and includes photographs of juveniles from several genera.

The Deuterostome Phyla

The following phyla are unique in their development and larval forms and, therefore, it is worthwhile to discuss some of their novel developmental characteristics before delving into the various larval forms. Two of these groups have received a great deal of study: the Echinodermata and the Chordata. Echinoderms have much in common with the chordates, for example both phyla exhibit radial cleavage. That is, after third cleavage, or at the eight-cell stage, the upper tier of four cells sits exactly over the lower tier of four cells. In most previously discussed taxa, the upper tier of four cells shifts from the radial to an interradial position. This twisting can be viewed as the beginning of a coil and is termed spiral cleavage. When third cleavage embryos of groups with radial cleavage (e.g., echinoderms) are viewed from the top, the upper tier is directly over the lower tier. Echinoderms have a coelom that is termed a **eucoelom** ("true coelom"), with the coelom forming from pouches in the **endoderm** of the gastrula. This pattern is typical of the chordates and echinoderms, while many of the animals exhibiting spiral cleavage form the coelom by splitting the gastrula's **mesodermal band**. The deuterostome phyla usually possess similar conditions in the categories discussed above. For this reason, the following three phyla are presented together.

The Lophophorates

The phyla with a whorl of feeding tentacles (called a **lophophore**) are collectively termed the lophophorates. Three lophophorate phyla are acknowledged: the Ectoprocta (Bryozoa), the Phoronida and the Brachiopoda. In benthic lophophorate adults, tentacles are ciliated to aid in directing food, born in currents of water in the ciliated channels, towards the mouth. The lophophore, or crown of tentacles, is characteristically urn-shaped. Some lophophorates are colonial (e.g., Bryozoans). Most of these tentacle-crowned worms have some sort of house they live in as adult (i.e., the bryozoan **zooecium**, phoronid tubes and brachiopod valves). All three of these phyla have larval stages which are planktonic.

Phylum Ectoprocta (Bryozoa)

One example of the benthic adult form of the bryozoans, or "moss-animals", is the roundish colony of *Membranipora*. This lophophorate grows, among other places, on the fronds of some large kelps. Like hydroids, branches and fragments of bryozoan colonies may occasionally be found in plankton tows. Adults collected from the side if docks or boats can also be interesting to observe (Figure 97.5). The flat, or encrusting, colonies, like *Membranipora*, often display a radiating pattern of the cell-like **zooid** chambers. Each chamber, or zooecium, is self-contained rather than interconnected as the colonial hydrozoans. Each new chamber is formed by asexual budding of the previous zooecium, hence the radiating pattern of the colony. A portion of a *Membranipora* colony, stripped from the kelp frond or found fragmented in the plankton, can be mounted on a slide and viewed under a compound microscope. Under these circumstances, movement and feeding of the colonies can be observed. The observer may also see differences in the young and old zooids (younger zooids are found towards the outside of the colony). In addition to asexual reproduction, bryozoans produce eggs and sperm.

In some bryozoans, including *Membranipora*, fertilized eggs develop into the planktonic **cyphonautes** larva (Figures 97.1-97.3). The cyphonautes larva resembles a cone that has been squashed flat, having two

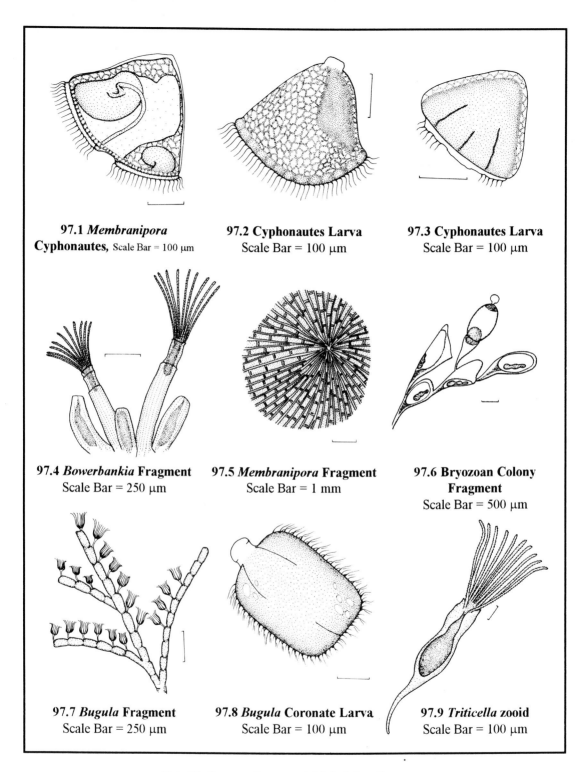

Plate 97. Bryozoan Larvae and Adult Colony Fragments

triangular valves. The ventral opening of the cone is lined with cilia for feeding and locomotion. An apical tuft of cilia aids navigation. When settling, the larva lies on its side and metamorphoses into an **ancestrula**, from which the rest of the colony will develop. With asexual budding of new individuals at the outer perimeter of the disc, the developing colony is soon a saucer of radiating zooids.

In other genera of Bryozoa (e.g., *Bugula*), a **coronate** larva develops from a ciliated gastrula (Figure 97.8). These larvae swim quickly, settle to the bottom and metamorphose into a new zooecium. If an embryo or coronate larva can be found in the plankton and isolated in the laboratory, this process of settlement and metamorphosis can be observed within a few days.

Some bryozoans build a series of brick-like chambers, others form chain-like colonies, with zooecia arranged along the chain in alternating or unilateral pairs. Fragments of *Bowerbankia* (Figure 97.4) taken in samples may have developing zooecia buds along the **stolon**. As with *Membranipora*, observation of laboratory-maintained individuals allows the researcher the opportunity to observe the development of the zooid (in buds along the stolon on *Bowerbankia*). When the zooid is fully developed, the zooecium opens and tentacles extend to begin collecting food. Forms with creeping stolons, such as *Bowerbankia*, can often be found growing on shells, rocks and ledges, both subtidally and in tidepools.

Ryland (1974) offers a review of the behavior, settlement and metamorphosis of the three classes of bryozoan larvae. This review, primarily descriptive text, is helpful for building an understanding bryozoan larval strategies. Photographs and illustrations of larvae from the class Gymnolaemata, including the coronate and cyphonautes forms, are available in Zimmer & Woollacott (1977).

Phylum Phoronida

One of the most beautiful of marine plankton is the **actinotroch** larva of the phylum Phoronida. These larvae gracefully swim in the plankton with hood upright, ciliated tentacles spread out canopy-like. The posterior end has a ciliary band, or telotroch, acting as a rear propeller for

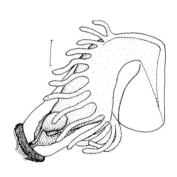

98.1 Actinotroch Larva
Scale Bar = 100 μm

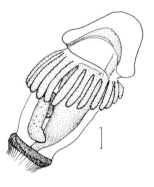

98.2 Actinotroch Larva
Scale Bar = 100 μm

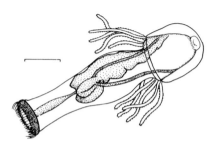

98.3 Actinotroch Larva
Scale Bar = 100 μm

98.4 Actinotroch Larva

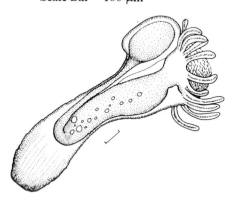

98.5 Late Actinotroch, Metamorphosis
Scale Bar = 100 μm

98.6 Juvenile Phoronid
Scale Bar = 100 μm

Plate 98. Phoronida: Actinotroch Larvae, Metamorphosis, and Juveniles

locomotion. The dramatic metamorphosis and settlement of an actinotroch to a benthic tube-dwelling adult begins as a bulging of the body wall near the mid-gut. This bulge grows until it is a significantly sized tube standing out from the main body. The hood disappears and tentacles emerge as the tube folds and brings the gut terminus near the anterior of the animal (resulting in the familiar U-shape of many tube-dwelling animals, Figure 98.5). Later, after a second metamorphosis, the animal straightens and resembles the benthic juvenile phoronid in Figure 98.6. This young lophophorate is ready to build a tube-home of sand grains and mucous secretions. The benthic adult of one genus, *Phoronis*, lives in a tube two to three mm in diameter and greater than 10 cm long. The worm, measuring about half of the tube's length, extends its lophophore to filter food from the seawater.

A thesis by Zimmer (1964) provides a good sampling of larval development in the phylum Phoronida. It includes discussions and descriptions of embryogenesis and larval stages in *Phoronopsis harmeri*, *Phoronis vancouverensis*, and *Phoronis pallida*. A review of phoronid biology (Emig, 1982), including information on planktonic larvae, provides an overview of the ecology of the phylum (e.g., settlement, feeding, reproduction, phylogenetics, etc.). Descriptions and illustrations of larval development in *Phoronis architecta* are given in a memoir by Brooks & Cowles (1905). This genus has a wide global distribution, though the species is Atlantic. As with many invertebrate phyla, congeneric phoronid species often exhibit similar ecology and development. It is useful to consult information available on related species.

Phylum Brachiopoda

Occasionally the fortunate plankton sample has in it a **lingula** larva (Figure 99.1) of the phylum Brachiopoda. The specimen illustrated was taken at the wharf in Santa Cruz, California. Brachiopod species which have this type of larva, in addition to a **lobate** larva (Brusca & Brusca, 1990), in their life-cycle are of the class Inarticulata. The other brachiopod class, the Articulata, have a lobate larva, but no lingulid. Some representative inarticulate brachiopods that bear a lingula larva (i.e.,

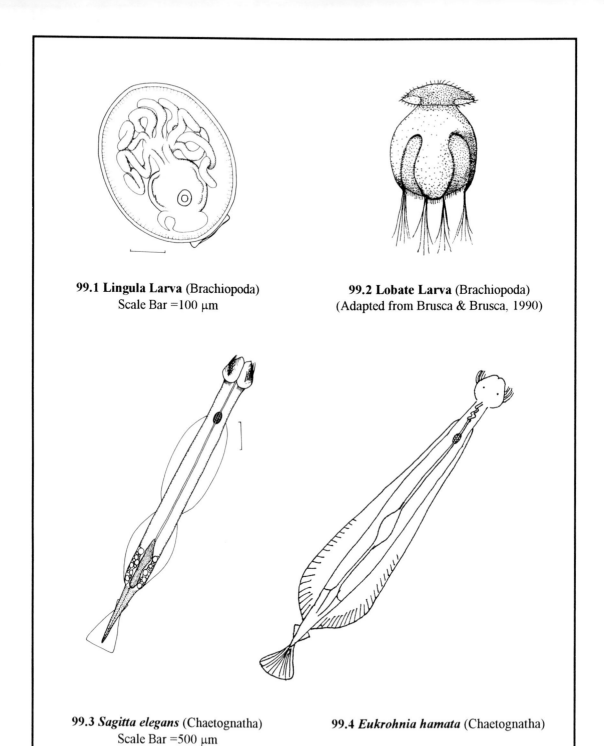

99.1 Lingula Larva (Brachiopoda)
Scale Bar =100 μm

99.2 Lobate Larva (Brachiopoda)
(Adapted from Brusca & Brusca, 1990)

99.3 *Sagitta elegans* (Chaetognatha)
Scale Bar =500 μm

99.4 *Eukrohnia hamata* (Chaetognatha)

Plate 99. Phylum Brachiopoda (Larvae) and Phylum Chaetognatha (Pelagic Adults)

possible species of the lingula in Figure 99.1) include *Crania, Discinisca, Glottidia* and *Lingula*. The lingula larva is morphologically very similar to the benthic adult brachiopod, with lophophores relatively large and protected within two valves. Notes, photographs and illustrations of larval development n various inarticulate brachiopods are available in Ashworth (1915), Chuang (1977) and Nielsen (1990). Settlement in this larva is simple as it contacts the substratum and the pedicle extends to attach.

Phylum Chaetognatha

Like the lophophorates, chaetognaths, or arrow worms, are deuterostomes. The arrow worm is a common component of coastal plankton, as well as oceanic plankton, or plankton farther from shore. It is nearly transparent and can easily be overlooked in a sample. When trying to locate arrow worms with a microscope, it is helpful to adjust the light source to different directions and intensities and search for sudden, swift stirring of particulate matter in the sample. Chaetognaths are a predaceous carnivore, which can sometimes be seen holding their prey in **grasping spines**, located in the anterior of the animal and surrounding the mouth. Arrow worms get their name from their long, slender bodies and their **lateral fins**, reminiscent of arrow feathers. Professor Smith has observed parasitic trematodes in the gut of some field-collected specimens. Chaetognaths are hermaphroditic or monoecious and may reciprocally exchange gametes with other individuals, though they can also self-fertilize. Egg-laden ovaries can be viewed through the transparent body wall. Fertilized eggs can be followed through early development if the animals are carefully maintained in the laboratory.

Arrow worms found in coastal waters of the North Pacific Coast are likely to be from the genus *Eukrohnia* or *Sagitta*. Lea (1955) and Fraser (1957) give more detailed descriptions of the different arrow worm species.

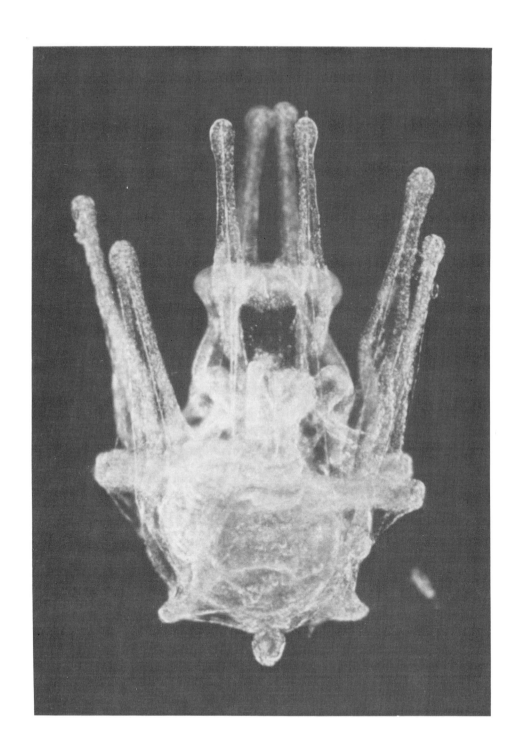

Phylum Echinodermata

Echinoderms, especially sea urchins and sea stars, have a long history of use in embryological research. Benthic adult representatives of this phylum are relatively abundant in most coastal habitats. Methods for obtaining gametes from adult echinoderms, as well as many other marine invertebrates, are described in M. Strathmann (1987).

Echinoderms are generally dioecious, having separate sexes. Some echinoderms (e.g., class Echinoidea, the urchins) can be induced to spawn gametes with potassium chloride injections. Spawned eggs can be kept for several hours and fertilized when the researcher is ready to observe. By placing unfertilized eggs on a microscope slide, the observer can add sperm as they watch through the microscope. Shortly after the sperm contact the egg, the **fertilization membrane** elevates above the egg's surface. This is seen through microscope as the appearance of a transparent halo around the egg. In the sand dollar *Dendraster excentricus,* development from fertilization through the blastula stage can be observed in a twenty-four hour period. The blastula is a hollow, somewhat spherical, embryological stage common to many animal phyla. In *Dendraster*, all of the development to this point takes place within the fertilization membrane and soon the blastula becomes ciliated. By ciliary action the blastula will rotate within the membrane. Later, the blastula dissolves the membrane chemically to become a free-swimming blastula. Eventually, due to the invagination of the blastopore, the blastula becomes a gastrula with a **primitive gut** formed from the endoderm. Development to this point is very similar in all of the echinoderm classes. However, beyond these early embryological stages, the development in different classes proceeds in different directions as the diverse larval forms take shape. Many of these interesting and beautiful larval forms occur in the plankton and may be captured in samples.

Most echinoderm larvae have cilia for feeding and locomotion. R. Strathmann (1971) analyzes the use of these ciliary bands in feeding and includes some line drawings of larvae of known identity from the phylum. Mortensen (1927) gives an overview of echinoderm larval form and illustrates the larvae of several species. Mortensen (1921) describes many echinoderm larvae.

Class Asteroidea (Sea Stars)

There are two primary larval forms in sea stars. The **Bipinnaria** larva bears a complex ciliated band which coils around the body. The **Brachiolaria** larva is also ciliated and bears long waving arms. In a late brachiolaria, one can see the developing juvenile rudiment with the secondary pseudo-radial symmetry of the adult form (e.g., Figure 101.5). Most echinoderm larvae, like asteroids, possess ciliary bands for feeding and locomotion. Development in one sea star from the second cleavage of the embryo all the way through to the newly metamorphosed juvenile is shown in Plates 100 & 101. Examples of descriptions of lecithotrophic asteroid larvae include Masterman (1972) and Birkeland et al. (1971). Development in a planktotrophic sea star, *Pycnopodia helianthoides*, is described in Greer (1962). The brachiolaria larva of *Pycnopodia helianthoides*, with its developing juvenile rudiment, is pictured on the cover of this guide.

Class Holothuroidea (Sea Cucumbers)

The sea cucumbers are represented in the plankton by several larval forms. One common holothuroid larva is the **auricularia** (Figures 102.2 & 102.3), which closely resembles the late bipinnaria larva of the sea stars. Holothuroids also have a larval stage known as a **doliolaria**. Professor Smith observed a doliolaria metamorphose into a juvenile sea cucumber (Figure 102.4).

Samples taken at Tomales Bay, California contained small, darkly pigmented balls floating free in the sample, rotating via ciliary action. They were isolated and the following day observed to have tentacular projections (Figure 102.8). A few days later it became clear these were **pentacula** larvae, which belong to the class Holothuroidea. These pentaculae were maintained in the laboratory until they metamorphosed into juvenile sea cucumbers (Figure 102.9), which were later identified as belonging to the

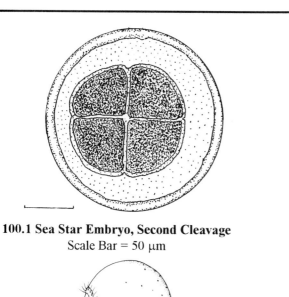

100.1 Sea Star Embryo, Second Cleavage
Scale Bar = 50 μm

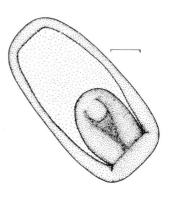

100.2 Sea Star Gastrula
Scale Bar = 50 μm

100.3 Early Bipinnaria Larva
Scale Bar = 50 μm

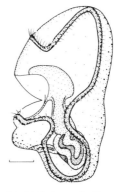

100.4 Bipinnaria Larva
Lateral View, Scale Bar = 100 μm

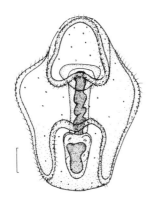

100.5 Bipinnaria Larva
Scale Bar = 50 μm

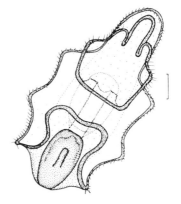

100.6 Late Bipinnaria Larva
Scale Bar = 100 μm

Plate 100. Echinodermata: Developmental Stages in Sea Stars (Class Asteroidea)

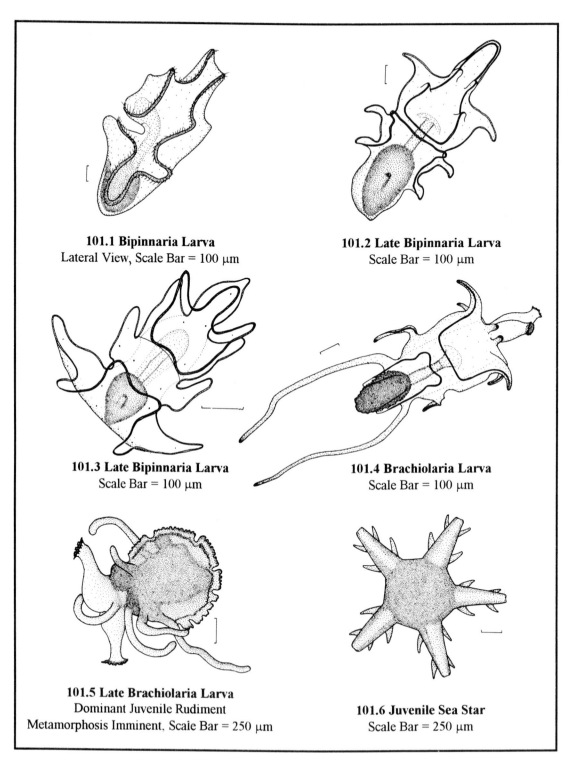

101.1 Bipinnaria Larva
Lateral View, Scale Bar = 100 μm

101.2 Late Bipinnaria Larva
Scale Bar = 100 μm

101.3 Late Bipinnaria Larva
Scale Bar = 100 μm

101.4 Brachiolaria Larva
Scale Bar = 100 μm

101.5 Late Brachiolaria Larva
Dominant Juvenile Rudiment
Metamorphosis Imminent, Scale Bar = 250 μm

101.6 Juvenile Sea Star
Scale Bar = 250 μm

Plate 101. Echinodermata: Developmental Stages in Sea Stars (Class Asteroidea)

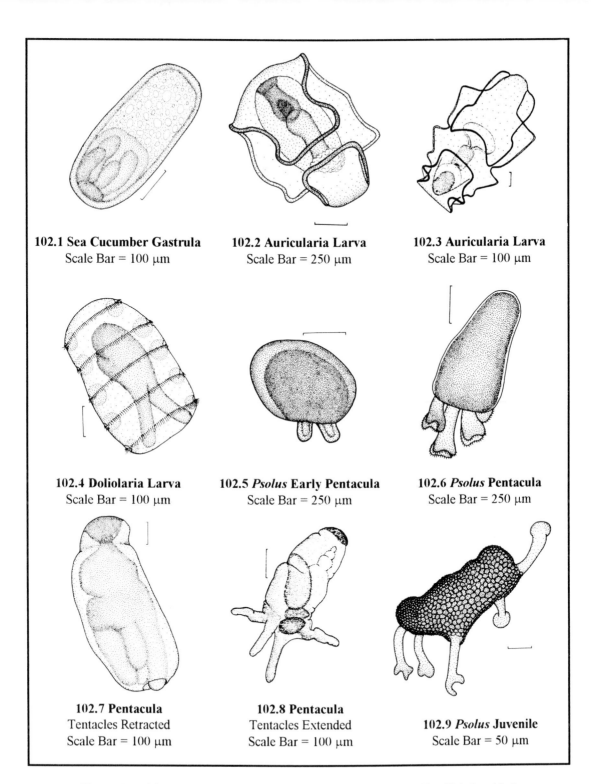

Plate 102. Echinodermata: Developmental Stages in Sea Cucumbers (Class Holothuroidea)

genus *Psolus*. Benthic adult *Psolus* have many epidermal skeletal plates. Like all adult echinoderms, sea cucumbers have a measure of **pentaradial symmetry**. Holothuroid larvae, especially the pentacula, bear a closer resemblance to the adult stage of their life-cycle than larvae from the other echinoderm classes. Examples of holothuroid development of two genera of sea cucumbers are available in Cameron (1985), McEuen & Chia (1985) and Smiley (1986a, 1986b).

Class Crinoidea (Feather Stars)

Not often seen in coastal plankton, crinoid larvae are probably found in deeper waters, as that is where many of the benthic adults are found. Crinoids also have a doliolaria larva, which exhibits the relation of this group to the other echinoderm classes by its similarity to the holothuroid doliolaria larva (Figure 102.4). Some literature is available with drawings and descriptions of crinoid larvae, though they probably do not provide sufficient information to identify an unknown larva to species. For example, Lacalli & West (1986, 1987) have published some photographs of larvae of the feather star *Florometra* in papers discussing ciliary band formation. Mladenov & Chia (1983) describes development in the same genus.

Class Echinoidea (Sea Urchins)

This class contains the sea urchins, which includes modified urchins such as the well-known sand dollars. The larvae in this class are called **pluteus** larvae. The pluteus is not exclusive to this class. Therefore, developing echinoids at this stage are termed an **echinopluteus**. Sea urchins are probably some of the most studied organisms by embryologists and developmental biologists. Developing larvae of many species have been successfully reared to settlement and metamorphosis. As with other classes and phyla, echinoids have some larvae that are planktotrophic, or plankton-feeding, and others that are lecithotrophic, or yolk-feeding. Lecithotrophic larvae can be reared to metamorphosis without having to bother with external food sources. Planktotrophic larvae (e.g., the sand dollar *Dendraster excentricus*) need some additional nutritional input to achieve metamorphosis, though early larval development (the first few days) usually proceeds without food. It should be noted that, though planktotrophic larvae reared without food may live for a while, they often exhibit abnormal morphology and will probably die before reaching metamorphosis. Many echinoplutei and newly metamorphosed echinoid juveniles are found in coastal plankton samples (e.g., Plate 103).

Information on the identification of echinoid larvae is surprisingly scarce considering the research that has been done on the physiology and embryology in the class. A publication containing descriptions of several genera is Vannucci (1961). R. Strathmann (1979) provides a key to echinoid larvae using skeletal rod characteristics. Photographs of an early pluteus of the genus *Allocentrotus* are displayed in Moore (1943).

103.1 *Dendraster* Echinopluteus

103.2 *Dendraster* Echinopluteus

103.3 Juvenile Urchin (Echinoidea)

103.4 Juvenile Urchin (Echinoidea)

Plate 103. Echinoid Larvae and Juveniles (Class Echinoidea)

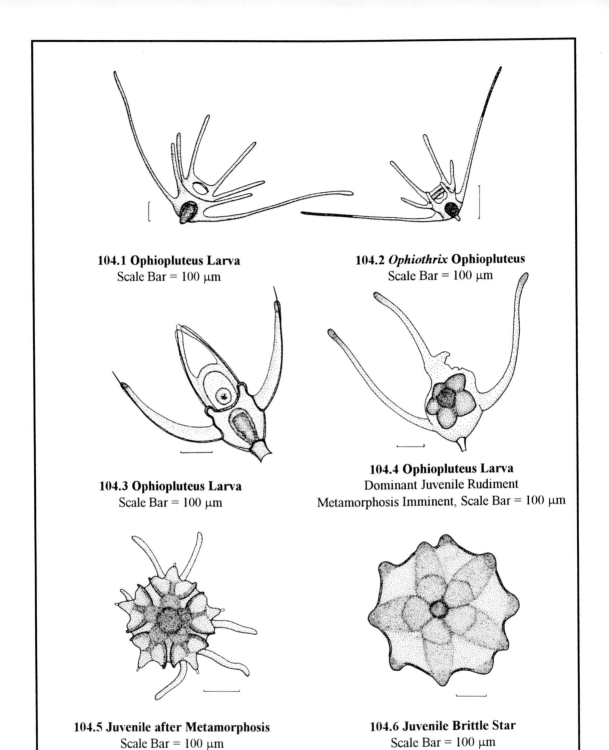

Plate 104. Echinodermata: Developmental Stages in Brittle Stars (Class Ophiuroidea)

Class Ophiuroidea (Brittle and Basket Stars)

Like echinoids, ophiuroids possess a pluteus larva, or a developmental form with elongate skeletal rods covered by a ciliated epidermis. In ophiuroids, the larva is termed an **ophiopluteus**. As in echinoplutei, the arms of the pluteus are lined with cilia which is used for moving smoothly and gracefully, as well as collecting food. Some ophiuroids spawn a planktonic larval form (e.g., Figures 104.1-104.4) while others are noted for brooding their young in the central disk's **bursal sac**. Brooding individuals release small benthic juveniles which crawl away to live relatively close to the adult. An ophiuroid brood can be followed through its development in the laboratory. For planktonic ophioplutei, it is possible to collect gametes or individuals caught in the plankton and rear them to metamorphosis in the laboratory.

Descriptions of larval stages in ophiuroids include illustrations of larvae from two genera (Thorson, 1934) and photographs of development in *Amphipholis* (Yamashita, 1985).

Phylum Hemichordata

Hemichordates are deuterostomes that possess a dorsal nerve cord. This group contains many planktonic forms including the larvae of pterobranchs and acorn worms. The typical hemichordate larva is the **tornaria**, reminiscent of a cross between mollusc trochophores and echinoderm auricularia. However, resemblance to other larvae is superficial and tornaria are unique unto themselves. Hemichordate larvae are generally rare in plankton samples from the North Pacific Coast, perhaps because adult hemichordates tend to live in warmer waters. Benthic adult *Glossobalanus* and *Saccoglossus* have been noted in certain intertidal areas, such as Franklin Point, North of Santa Cruz, California, where Professor Smith and his marine biology class located some specimens. These tornaria larvae (e.g., Figure 105.1) were collected over several years in plankton tows in Central California waters.

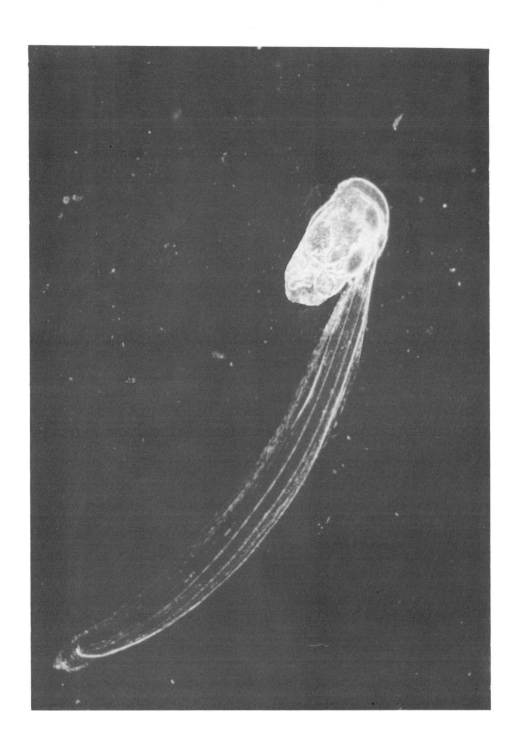

Phylum Chordata

The phylum Chordata has three subphyla, all of which are represented in the plankton. Two of the subphyla are invertebrate phyla, the Cephalochordata and the Urochordata. Cephalochordates, or lancelets, are usually benthic and live in the sand, while they have larvae that are planktonic for several months. Lancelet larvae are not often found in significant abundances in the plankton. The urochordates, or tunicates, have many representatives in the plankton, both larval and adult forms. There are four classes of tunicates: Ascidiacea, Appendicularia, Thaliacea and Sorberacea. Sorberaceans are deep-sea organisms and are not regularly found in coastal plankton samples. Adult appendicularians and salps (class Thaliacea) are marine planktonic organisms. Many Ascidians (sea squirts) have larvae that are planktonic for hours to days. A good source for learning about the various tunicates, including some line drawings with corresponding identities, is Berrill (1950). The third chordate subphylum is the Vertebrata, including the fishes. Many fishes have planktonic larvae, some of which are illustrated in this guide.

Class Ascidiacea (Subphylum Urochordata)

The ascidian (sea squirt) form most often encountered in the plankton is the **tadpole larva**. Rarely will the adult form, which is sessile and benthic, be found in plankton tows, though occasionally they may be seen settled on detached debris. Typical adults, which are benthic, possess incurrent and excurrent **siphons**. Through these siphons, water is transported by the ciliary action of gills in the **branchial chamber**. These animals are enclosed in a gelatinous **tunic**, hence the name "tunicate". Isolated tadpole larvae will settle, lose their tail and metamorphose to the adult stage. Some specimens have transparent body walls and the ciliary action of the gills and the flow of the circulatory system can be viewed under the microscope. Various settled ascidians are shown in Plate 105. Larval tadpoles taken in samples from the Central California coast are shown in Plate 106. These tadpole larvae exhibit some diversity within their basic form. The heads of the larvae possess adhesive **papillae**, used

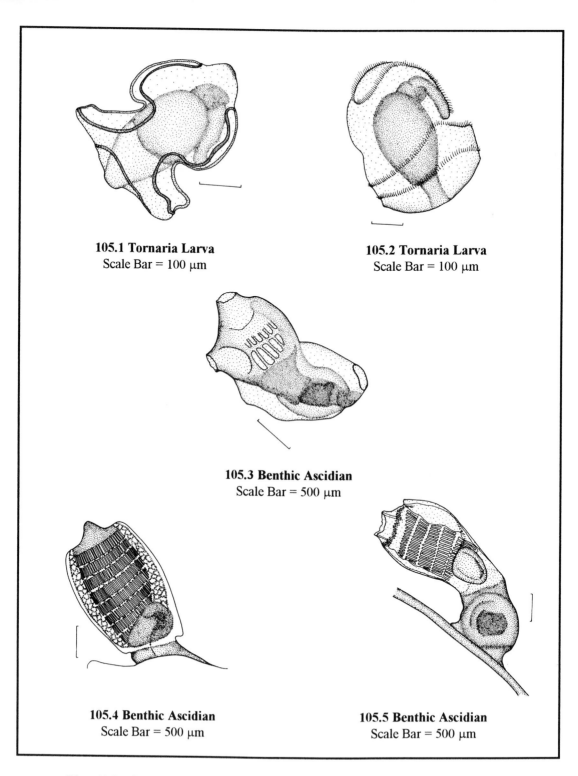

Plate 105. Phylum Hemichordata (Larvae) and Phylum Chordata (Class Ascidiacea)

to secure the tadpole to the substratum when metamorphosis begins. Common to all is the **notochord**, a typical chordate structure that regresses as metamorphosis proceeds in ascidians. Once the planktonic larva has settled and begins the process of metamorphosis, the **resorption** of the tail is quite rapid. Several tadpoles were observed to lose half of their tail in just eight hours. Figure 106.4, believed to be the larva of *Metandrocarpa*, shows the larva to possess a number of finger-like projections. These projections are the **ampullae** for attachment of the post-metamorphosis juvenile. When settling, the ampullae form a radiating basal attachment of finger-like processes interconnected within the young colony. Patterns of flow and reversal of flow in the circulatory system can be viewed through the body wall. Tadpole larvae kept in the laboratory will soon settle, metamorphose and begin forming a new colony. During colony formation, new buds are produced asexually by the primary ascidian.

Larval structure in the ascidian *Styela* is illustrated in Grave (1944). Other tadpole larvae are sometimes described in literature where the main thrust is the behavior, evolution or physiology of the larval form. For instance, the larva of *Pyura* is shown in Svane (1982) and a photograph of a *Diplosoma* tadpole is displayed in Cavey & Cloney (1976).

Class Appendicularia (Subphylum Urochordata)

Representatives of the class Appendicularia, also known as larvaceans, are shown in Plate 107. These tunicates retain the notochord and tail in the adult form, having no metamorphosis comparable to that of ascidians. Because of the resemblance of ascidian tadpole larvae and adult appendicularians, evolutionary biologists believe the appendicularian **clade** arose evolutionarily from an ancestral form with a tadpole larva by a process called **paedomorphosis**. Paedomorphosis is a term applied to speciation that occurs when a clade loses the adult stage and reproduces in the form that belonged to the ancestral larva. Common genera from the North Pacific include *Oikopleura* and *Fritillaria*. *Oikopleura* lives within a gelatinous house, which they secrete around their body. Hamner (1974) provides photographs, taken *in situ*, of these animals with intact houses. Within this house they produce a mucous feeding net to filter out food

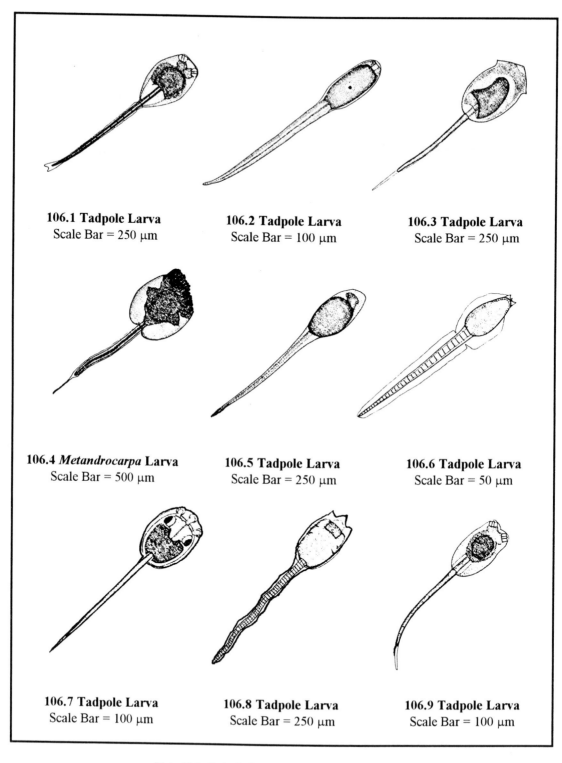

Plate 106. Tadpole Larvae (Subphylum Urochordata)

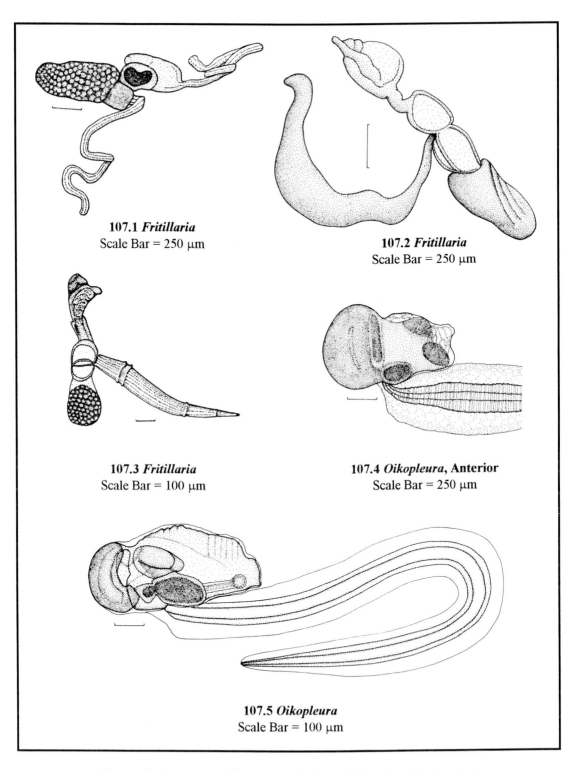

Plate 107. Larvaceans (Class Appendicularia, Subphylum Urochordata)

from the seawater which is pumped through the gelatinous housing. This spherical house is usually destroyed during sampling unless special measures are taken to obtain specimens intact. One way to collect larvaceans with their house intact is to visually locate the individual from a dock or boat and then gently dip them from the ocean's surface in a jar. Many larvaceans are large enough to see with the naked eye (1-2 mm in length), though most remain small enough that locating them from a dock can be difficult at first. It is wise to collect a plankton sample with a net to confirm the presence of the appendicularians before spending much time staring over the side of a dock. Most appendicularians have a distinct rapid jerking motion which will help them to be located in net samples, though they are often largely transparent. Professor Smith has observed that *Fritillaria* is often found swimming or floating at the top of a plankton sample (collected by a traditional tow of a plankton net), while *Oikopleura* is often found at the bottom of the sample unless actively swimming. In addition to the Berrill (1944) reference, Bückmann (1945, 1969) and Bückmann & Kapp (1975) present keys for identifying appendicularian genera that are global in distribution. Fenaux (1967) describes larvacean genera from the Mediterranean and seas near Europe. Galt (1970) has some illustrations of some of the larvaceans found in the Puget Sound.

Class Thaliacea (Subphylum Urochordata)

Thaliaceans, commonly called salps, are floating tunicates. Most are cylindrical, tubular or barrel-shaped, as in *Doliolum* (Figure 108.4). Some other salps are illustrated in Plate 108. Though many, such as the doliolids, are as small as one centimeter, certain salps form colonies meters long and some individuals are reported to approach this size singly. Being transparent and large, living salps provide a remarkable display of ciliary action in the **pharyngeal basket** and pumping and flow in the circulatory system. In addition to Berrill (1944), mentioned above, Fraser (1947a, 1947b) describe tunicates from the families Salpidae and Doliolidae.

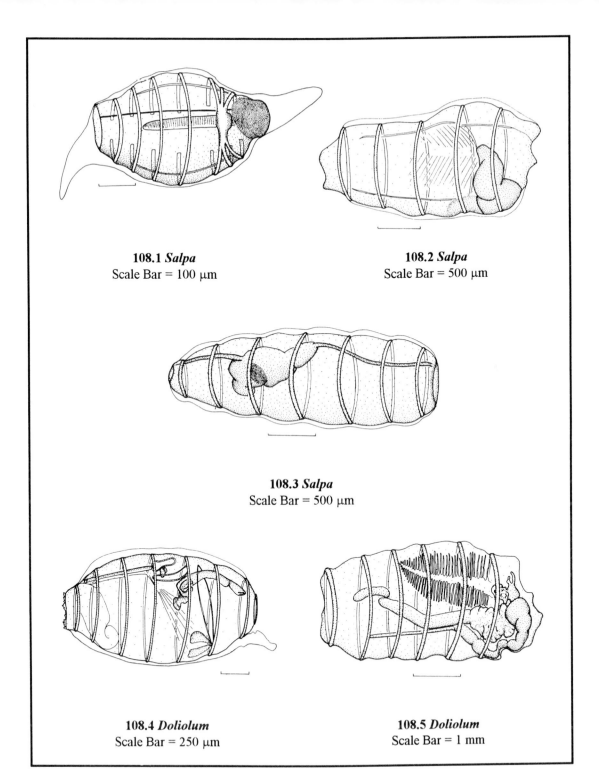

Plate 108. Salps (Class Thaliacea, Subphylum Urochordata)

Subphylum Vertebrata (Phylum Chordata)

The vertebrates are represented in the plankton by the eggs and newly hatched larvae of many fish species. These fish larvae are regularly caught in plankton tows. Larval fish populations are usually at a relatively low density. When collecting fish larvae, it is generally recommended to tow for a longer period of time using a larger mesh and capacity net (e.g., a 200 µm mesh net 1 m in diameter). Occasionally, fishes may be caught that are unexpectedly large. For example the pipefish illustrated in Figure 112.3 was nearly four centimeters in length. Whether or not fishes like this are truly planktonic, or simply captured incidentally, is an item for debate. For most fishes there comes a time when larvae settle out of the plankton. However, many adult forms are not benthic or **demersal**, but nektonic. Planktonic larvae of these nektonic species "settle" out of the plankton, but continue living in the water column and join the swimming adult population.

Several eggs and their respective hatchlings are illustrated in Plates 109 & 110. In Central California, anchovy eggs and larvae are periodically (seasonally) abundant. Larvae captured in tows can be observed in the laboratory. The internal organs of planktonic fishes are often distinguishable through the transparent body wall. Eyes, nerve cord, gill slits, heart and **myotomes**, or segmental muscle bundles, can be seen clearly as the day to day development of the larva is observed. Planktonic fish larvae often possess yolk reserves (e.g., Figure 110.3) intended to help them survive an awkward period of incompetence and vulnerability in early life. This ventral **yolk sac** atrophies as the young fish grows and develops. Finally, the yolk sac disappears and the fish must forage for food.

Pleuronichthys coenosus possesses an unusual egg (Figure 110.4). The membrane of the egg is modified with a honeycomb-like sculptured chorion. While in the egg, the young larval form becomes heavily pigmented (Figure 110.3), finally having many chromatophores as it hatches from the egg to swim freely in the plankton (Figure 110.5). Some other larval fishes captured in plankton hauls are illustrated in Plates 111 & 112. Matarese et al. (1989) is recommended for more assistance with identifying early developmental stages of fishes from the Northeast Pacific.

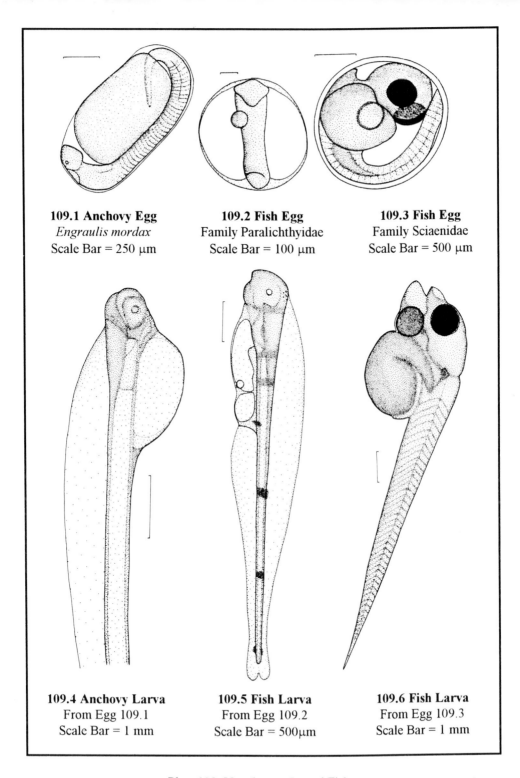

Plate 109. Vertebrata: Larval Fishes

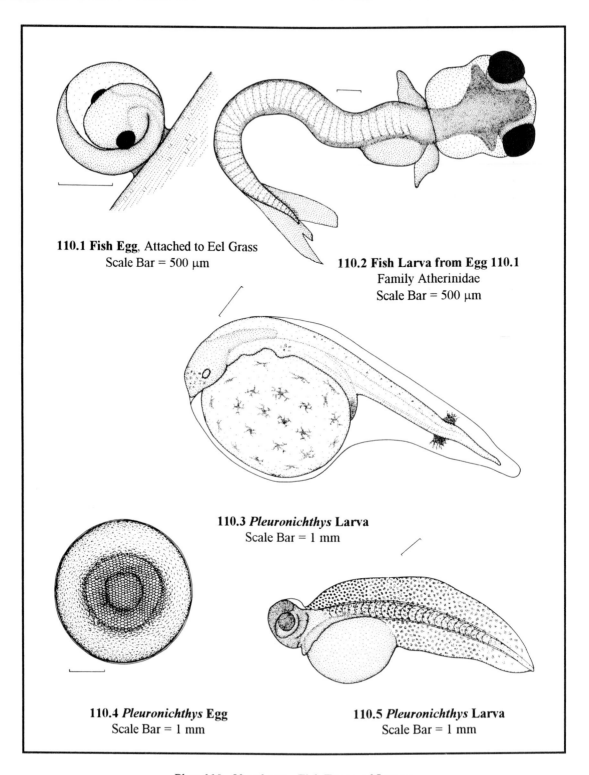

Plate 110. Vertebrata: Fish Eggs and Larvae

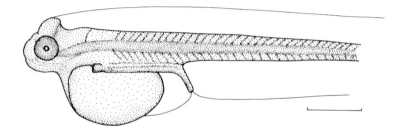

111.1 Fish Larva Family Pleuronectidae (Scale Bar = 500 μm)

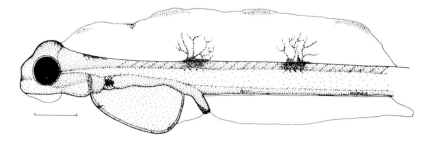

111.2 Fish Larva Family Pleuronectidae (Scale Bar = 500 μm)

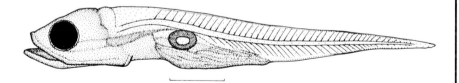

111.3 Fish Larva Family Myctophidae (Scale Bar = 500 μm)

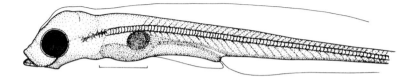

111.4 Fish Larva Family Myctophidae (Scale Bar = 500 μm)

Plate 111. Vertebrata: Larval Fishes

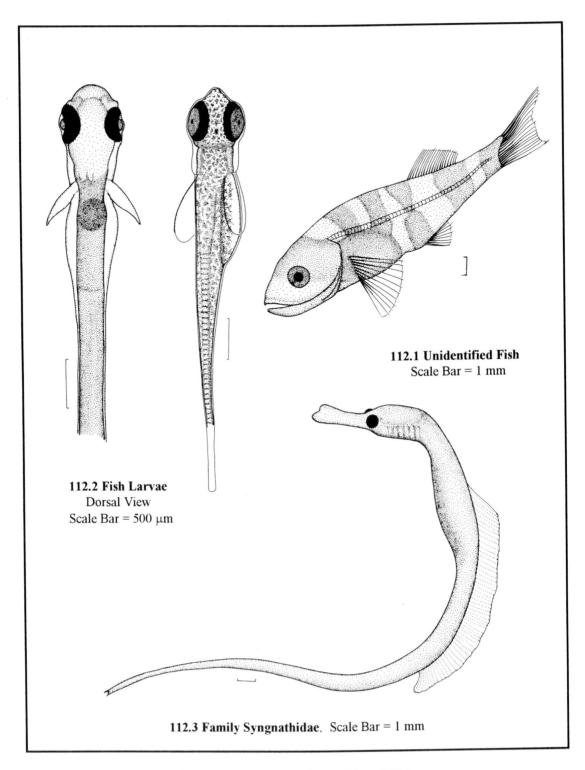

112.1 Unidentified Fish
Scale Bar = 1 mm

112.2 Fish Larvae
Dorsal View
Scale Bar = 500 μm

112.3 Family Syngnathidae, Scale Bar = 1 mm

Plate 112. Vertebrata: Fishes and Larval Fishes

Glossary

Actinotroch - Larval stage in many phoronids (phylum Phoronida)
Actinula - Larval stage in some hydrozoans (phylum Cnidaria)
Adductor Muscles - In larval and adult bivalves (phylum Mollusca), muscles used to hold the valves closed
Amphiblastula - Larval stage in some sponges (phylum Porifera)
Ampullae - Adhesive projections, for attachment of settled ascidians (phylum Chordata)
Ancestrula - Original settled zooid in a bryozoan colony
Aphotic Zone - Ocean's depths where light attenuation is complete, usually > 1-200 m
Apical Tuft - Ciliary cluster for sensory reception found on the anterior of many larvae
Auricularia - Larval stage in some sea cucumbers (phylum Echinodermata)
Autotrophic - Having the ability to manufacture energy and grow using only raw materials
Auxospore - Reproductive structure in diatoms
Axopodia - Cytoplasmic extensions on skeletal axes, Phylum Actinopoda
Bell - The hood or umbrella, often dome-shaped, of a medusa (phylum Cnidaria)
Benthic - Of or pertaining to the benthos, the sea floor or bottom of the ocean
Biramous - Branching into two parts, forking
Biomass - The total volume of life in a system, regardless of the number of individuals
Bipinnaria - Larval stage in some asteroids (phylum Echinodermata)
Blastopore - Hole created by the invagination of cells in the blastula sphere
Blastula - Early developmental stage in many animals, a hollow sphere of cells
Bloom - Population explosion phenomenon in the plankton, most often referring to phytoplankton
Brachiolaria - Larval stage in asteroids (phylum Echinodermata)
Bract - Shield-like zooid helping to make-up a siphonophore cormidium
Branchial Chamber - Cavities where respiration takes place, often where gills are found
Brood - To provide protection within a pouch or cavity for young which have hatched
Budding - Asexual clone production via the outgrowth and separation of tissue
Bursal Sac - Ventral pouch in adult ophiuroids (phylum Echinodermata) used to brood
Calymma - Actinopod cytoplasm located outside the central capsule
Calyptoblast - Thecate hydrozoa, as in the leptomedusae
Central Capsule - The center of the actinopod skeletal network
Cephalization - The tendency of animals to organize and control from the head region
Cephalothorax - Fused head and thorax
Ciliate - A phylum of eukaryotic microorganisms bearing cilia
Chaetae - Spiny projections in polychaete worms, often arising from parapodia
Chela - Clawed appendage in crustaceans

Glossary

Chloroplast - Organelle in photoautotrophic cells where photosynthesis takes place
Chorion - Protective membrane surrounding an egg
Chromatophores - Pigment cells
Clade - An evolutionary lineage or monophyletic group
Cod-End Bucket - Terminal container of a plankton net, for collection of the sample
Coeloblastula - Specific type of blastula which has a hollow cavity
Coelom - A hollow cavity between the gut and outer body wall
Colloblasts - Specialized adhesive cells in ctenophores
Colonial - Conspecific individuals which live in close proximity and may rely upon one another for survival
Consumer - Organism which grazes exogenous food for energy and growth
Corbula - Planula brood chamber (phylum Cnidaria)
Cormidia - Subcolonies budding along the stem in calycophorans (order Siphonophora)
Corona - Crown of cilia at the head of rotifers
Coronate Larva - Larval stage in some bryozoans
Cryptoniscid - Larval form of some parasitic isopods (phylum Arthropoda)
Ctenes - Fused cilia paddles, giving the phylum Ctenophora its name
Cydippid - Larval stage in comb jellies (phylum Ctenophora)
Cyphonautes - Larval stage in some bryozoans
Cyprid - Larval stage in barnacles
Cytoplasm - Extranuclear cellular protoplasm
Demersal - Living near or on the bottom
Deuterostomes - Organisms in which the blastopore becomes the anus
Diatoms - Golden-brown algae, responsible for much primary production at the base of the ocean's food web
Dioecious - Having sexes in separate individuals
Direct Development - Progressing from a zygote to the adult form with no intermediate larval stage
Doliolaria - Larval stage in some holothuroids (phylum Echinodermata)
Echinopluteus - Pluteus larva in echinoids (phylum Echinodermata)
Echinospira - Larval stage in some gastropods
Endoderm - Innermost germ cell layer
Endopods - Inner rami of biramous appendages in arthropods
Ephyra - Medusa-like larval stage in cnidarians (phylum Cnidaria)
Epitheca - Outer valve or frustule in diatoms
Epitoke - Posterior reproductive region in some polychaetes
Eucoelom - Having a mesodermally derived coelom ("true coelom")
Eudoxid - Independently living cormidium fragment (order Siphonophora)
Eukaryotic - Having a cell or cells with a nucleus and organelles

Glossary

Euphotic Zone - The upper 100-200 m of the ocean where light penetrates and photosynthesis can take place
Extant - Still living, not extinct
Exumbrella - Outer surface of the dome or umbrella of a medusa (phylum Cnidaria)
Fertilization Membrane - Thin covering arising from an egg upon fertilization
Flagellate - Grouping of protoctists which bear one or more flagella
Flagellum - Locomotory structure in flagellates, similar to cilia, but larger and occurring individually or in pairs
Flow Meter - Device for monitoring rate of flow, often attached to plankton nets
Foraminiferan - Shelled ameba with calcareous skeleton
Foraminiferan Ooze - An accumulation of viscous seafloor sediments comprised of calcareous foraminiferan skeletons
Fusiform - Torpedo-shaped or streamlined
Gastrovascular Cavity - Digestive cavity in cnidarians
Gastrozooid - Feeding zooid in hydroid polyps
Gastrula - Embryological stage succeeding the blastula stage
Gnathopod - Crustacean appendage
Gonophores - Medusa buds produced in gonozooids (class Hydrozoa)
Gonotheca - Cup-like extension of the perisarc covering hydrozoan reproductive polyps
Grasping Spines - Sharp projections surrounding the mouth of chaetognaths
Gymnoblasts - The athecate hydrozoa, also called anthomedusae
Hermaphroditic - See Monoecious
Heterotrophic - Having the ability consume exogenous organic matter as well as the ability to produce energy from raw materials
Hinge - Interlocking and attached portion of paired bivalve shells
Holoplankton - Organisms living an exclusively planktonic existence
Homologous - Having a similar structural derivative in the ancestor
Hydromedusa - Medusa stage in hydrozoans (phylum Cnidaria)
Hydrotheca - Cup-like extension of perisarc covering hydrozoan feeding polyps
Hypotheca - Inner valve or frustule in diatoms
Indirect Development - To pass through a larval stage during progression from zygote to adult
Infaunal - living or burrowing in sediments such as sand or mud
Interradial - Between primary canals or septae in cnidarians
Isogametes - Reproductive spores, eggs or sperm, of equal size
Lateral Fins - Stability rudders on fish and chaetognaths
Lecithotrophic - Yolk-feeding, needing no exogenous food to achieve metamorphosis
Lingula - Larval form in some brachiopods
Littoral Zone - Pertaining to the coast, intertidal
Lobate Larva - Generalized term for larvae of brachiopods
Longitudinal Muscles - Muscles running lengthwise

Glossary

Lophophore - Crown of tentacles characterizing the lophophorate phyla (Bryozoa, Brachiopoda, and Phoronida)
Lorica - Cup-like outer casing or house for tintinnids and related ciliates
Manubrium - Tubular support for mouth in medusae, hangs from subumbrellar surface
Margin - Outside edge, referring to outer edge of the medusa bell in cnidarians
Medusa - Umbrella-shaped swimming stage in hydrozoans and scyphozoans (phylum Cnidaria), often called "jellyfish"
Megalopa - Late larval stage in many decapods, pre-metamorphosis stage
Meridonal - Top to bottom on the outer surface of a comb jelly sphere
Meroplankton - Organisms planktonic for only a portion of their life-cycle
Mesodermal Band - Middle germ cell layer
Metamorphosis - Dramatic change in form as a larva changes into an adult
Metatrochophore - Advanced polychaete trochophore with many segments
Microbial Food Web - The cycling and flow of nutrients by phytoplankton and bacteria
Microniscid - Larval form of some parasitic isopods (phylum Arthropoda)
Mitraria - Larval form in the polychaete *Owenia* sp.
Molecular Systematics - The determination of phylogenetic relationships using similarities of DNA, enzymes, and other molecules
Monoecious - When individuals possess both male and female sexual organs, hermaphroditic
Müller's Larva - Larval stage in some flatworms (class Turbellaria)
Myotomes - Repetitive muscular segments
Mysis-Zoea - Larval stage in some decapods resembling mysid shrimp
Naupliar Eye - Optic sensory organ in nauplius larvae
Nauplius - Larval stage in many crustaceans
Nectochaete - Larval stage in nereid polychaetes (phylum Annelida)
Nectophores - Siphonophore swimming bell or zooid
Nektonic - Of or pertaining to a life of swimming in which regional distributions are determined by the organisms themselves rather than by ocean currents
Nematocysts - Stinging cells in cnidarians
Neritic - Zone of coastal waters, or pertaining to coastal life
Neurotroch - Midventral ciliary band in many larvae
Notochord - Skeletal rod in chordates, lost in most adult tunicates
Oblique - Having a wide angle, term applied to plankton nets hauled behind a boat
Oogonium - Egg chamber, as in the female hydroid gonangium
Ophiopluteus - Pluteus of ophiuroids (phylum Echinodermata) with elongate arms
Paedomorphosis - When sexually mature adults display larval characters from related species, a type of heterochrony
Papillae - Adhesive structures in ascidian tunicates
Parapodia - Lobe-like appendages in polychaetes

Glossary

Parenchymula - Larval stage in some sponges (phylum Porifera)
Pediveliger - Late larval stage in bivalves (phylum Mollusca), with foot for settlement
Pelagic - Zone of the open ocean, or pertaining to oceanic life
Pelagosphaera - Larval stage in some sipunculids
Pentacula - Larval stage in holothuroids (phylum Echinodermata)
Pentaradial Symmetry - Pseudo-radial symmetry of adult echinoderms
Periopods - Crustacean appendages
Perisarc - Chitinous exoskeleton secreted by the epidermis in hydroids
Pharyngeal Basket - Ciliated structure by which feeding occurs in urochordates
Photosynthesize - To produce energy and growth from light and raw nutrients
Phylogenetic Relationships - Evolutionary affinities of groups with common traits
Phytoplankton - Generic term for photosynthesizing plankters (diatoms, etc.)
Pilidium - Larval stage in some nemerteans
Planktonic - Floating in water, having a regional distribution determined by currents
Planktotrophic - Larvae that need to ingest exogenous food while in the plankton in order to achieve metamorphosis
Planula - Larval stage in some cnidarians (phylum Cnidaria)
Pleopods - Crustacean appendages
Pluteus - Larval stage in some echinoderms
Polyp - Sessile, erect, "hydra-like" stage in the life-cycles of cnidarians
Polytroch - Advanced polychaete trochophore larval with several or more segments
Primary Production - Initial growth and energy created by autotrophs
Primitive Gut - "Archenteron", cavity in embryos to become digestive gut in adults
Protista - Eukaryotic microorganisms (not including animals, plants and fungi)
Protostomes - Organisms in which the blastopore becomes the mouth
Prototroch - Most anterior ciliary band in many larvae
Protozoea - Early larval stage in decapods, also called "prezoea"
Pygidium - Posterior segment in polychaetes (phylum Annelida), origin of new segments
Radial - On the radius, such as the primary canals or septae in cnidarians. Refers to symmetry
Radial Canals - Branches of the gastrovascular cavity in medusae
Radial Cleavage - When the top tier of cells are directly above the bottom tier at the 8-cell stage (after the third cleavage)
Radiolarian - Planktonic actinopod protoctists, dead skeletons of which form siliceous oozes
Radiolarian Ooze - Viscous seafloor sediment comprised of siliceous radiolarian tests
Resorption - To dissolve and assimilate
Reticulopodia - Complex network of pseudopodial cytoplasm extensions
Ring Canal - Marginal connecting canal in medusae, opens into radial canals

Glossary

Rostrum - Anterior projection of cephalothorax, usually between eyes
Scyphistoma - Attached polyp stage of scyphozoans (phylum Cnidaria)
Septae - Wall-like partitions in a cavity (e.g., in adult sea anemones)
Sessile - Staying in one place, being attached
Siphons - Structures for directing water currents, used for feeding and respiration
Somites - Body segments
Spicules - Sponge structural network
Spiral Cleavage - When the top tier of cells are offset from the bottom tier at the 8-cell stage (after the third cleavage)
Statocysts - Balance organs
Stem - Extension of swimming bell in calycophorans from which cormidia bud
Stolon - Root-like filament for attachment and propagation
Strobilization - The process of ephyra production by a scyphistoma
Styliform - Style shaped, bristle-like
Subumbrella - The underside, or concave surface, of the dome of a medusa
Subumbrellar Cavity - Hollow space underneath the dome of a medusa
Symbiotic - Living together
Tadpole Larva - Larva of ascidians (phylum Chordata)
Tagmosis - The accumulation of specialized functions in certain regions
Teloblastic - Tissue growth in a chain-like formation
Telotroch - Most posterior ciliary band in many larvae
Telson - Terminal segment in crustaceans
Tentacle - Specialized organ, often slender and prehensile, for food gathering
Test - Shell or valve, skeletal structure
Tornaria - Larva of some hemichordates
Trochophore - Early larval stage in many polychaetes and molluscs
Trophic - Of or pertaining to energy or the production or acquisition of energy
Tunic - Outer transparent celluloid covering of the tunicates (subphylum Urochordata)
Umbo - Ridge or knoll on bivalve shell
Upwelling - The rise of deep, cold, nutrient-rich water into the euphotic zone
Uropods - Crustacean appendages, form a tail fan in combination with the telson
Urosome - Most posterior body region
Veliger - Larval stage in gastropods (phylum Mollusca)
Velum - Ciliated extension of the mantle in veliger larvae
Yolk Sac - Membranous sac attached to young fishes and containing yolk for nutrition
Zooecium - Individual chamber, or "house", in bryozoans
Zooid - Individual member of a colony
Zooplankton - Plankton belonging to the Kingdom Animalia
Zygote - A fertilized egg

References

Ally, J. R. R. (1974). A description of the laboratory-reared first and second zoeae of *Portunus xantusii* (Stimpson) (brachyura, decapoda). Calif. Fish and Game 60 (2): 74-78.

Ally, J. R. R. (1975). A description of the laboratory-reared larvae of *Cancer gracilis* Dana, 1852 (decapoda, brachyura). Crustaceana 28(3): 231-246.

Amio, M. (1963). A comparative embryology of marine gastropods, with ecological considerations. The Journal of the Shimonoseki University of Fisheries 12: 229-358.

Andersen, R. A., D. M. Jacobson and J.P. Sexton (1991). Catalog of Strains. West Boothbay Harbor, Provasoli-Guillard Center for Culture of Marine Phytoplankton.

Anger, K., M. Montu, C. DeBakker and L.L. Fernandes (1990). Larval development of *Uca thayeri* Rathbun, 1900 (decapoda: ocypodidae) reared in the laboratory. Meeresforsch. 32: 276-294.

Arai, M. N. and A. Brinckmann-Voss (1980). Hydromedusae of British Columbia and Puget Sound. Canadian Bulletin of Fisheries and Aquatic Sciences Bulletin 204 (Ottawa): 192.

Ashworth, J. H. (1915). On the larvae of *Lingula* and *Pelagodiscus* (Discinisca). Transactions of the Royal Society of Edinburgh 51: 45-69.

Banner, A. H. (1950). A taxonomic study of the Mysidacea and Euphausiacea (Crustacea) of the northeastern Pacific. Part III. Euphausiacea. Trans. Royal Canad. Inst. 28: 1-63.

Barnard, J. L. (1969). The families and genera of marine gammaridean Amphipoda (Crustacea). U.S. Nat. Mus. Bull. 271. 535 pp.

Barnes, H. and M. Barnes (1959). The naupliar stages of *Balanus nubilis* Darwin. Canadian Journal of Zoology 37: 15-23.

Berrill, N. J. (1950). The Tunicata with an account of the British species. London, Quaritch.

Bhaud, M. and C. Cazaux (1987). Description and identification of polychaete larvae; their implications in current biological problems. Oceanis 13(6): 596-753.

Bidwell, J. P. and S. Spotte (1985). Artificial Seawaters: Formulas and Methods. Boston, Jones & Bartlett Publ., Inc.

Birkeland, C., F. S. Chia and R.R. Strathmann (1971). Development, substratum selection, delay of metamorphosis and growth in the seastar, *Mediaster aequalis* Stimpson. Biological Bulletin 141: 99-108.

References

Blake, J. A. (1975a). The larval development of Polychaeta from the Northern California coast. II. *Nothria elegans* (Family Onuphidae). Ophelia 13: 43-61.

Blake, J. A. (1975b). The larval development of Polychaeta from the Northern California coast. III Eighteen species of errantia. Ophelia 14: 23-84.

Boletzky, S. (1974). The "larvae" of cephalopoda: a review. Thalassia Jugoslavica 10(1/2): 45-76.

Bousquette, G. D. (1980). The larval development of *Pinnixa longipes* (Lockington, 1877) (Brachyura: Pinnotheridae), reared in the laboratory. Biological Bulletin 159: 592-605.

Bowman, T. E. and H. E. Gruber (1973). The families and genera of Hyperiidea (Crustacea: Amphipoda). Smith. Contr. Zool. (146): iv + 64 pp.

Branscomb, E. S. and K. Vedder (1982). A description of the naupliar stages of the barnacles *Balanus glandula* Darwin, *Balanus cariosus* Pallas, and *Balanus crenatus* Bruguiere (Cirripedia, Thoracica). Crustaceana 42(1): 83-95.

Brooks, W. K. and R. P. Cowles (1905). *Phoronis architecta:* its life history, anatomy, and breeding habits. Memoirs of the National Academy of Sciences 10: 72-113.

Brown, S. K. and J. Roughgarden (1985). Growth, morphology, and laboratory culture of larvae of *Balanus glandula* (Cirripedia: Thoracica). Journal of Crustacean Biology 5(4): 574-590.

Brusca, R. C. and G. J. Brusca (1990). Invertebrates. Sunderland, Sinauer Associates, Inc.

Bückmann, A. (1945). Appendicularia I-III. Conseil international pour l'exploration de la mer Zooplankton sheet 7: 1-8.

Bückmann, A. (1969). Appendicularia. Conseil international pour l'exploration de la mer Zooplankton No. 7: 1-9.

Bückmann, A. and H. Kapp (1975). Taxonomic characters used for the identification of species of Appendicularia. Mitt. aus dem Hamburgischen Zoologischen Mus. und Institut 72: 201-228.

Cameron, J. L. (1985). Reproduction, development, processes of feeding and notes on the early life history of the sea cucumber *Parastichopus californicus* (Stimpson). Ph.D. Thesis, Simon Fraser University.

Cavanaugh, G. M. (1975). Formulae and Methods of the Mrine Biologyical Laboratory Chemical Room, 6th edition. Woods Hole, Marine Biological Laboratory.

Cavey, M. J. and R. A. Cloney (1976). Ultrastructure and differentiation of ascidian muscle. Cell and Tissue Research 174: 289-313.

References

Chanley, P. and J. D. Andrews (1971). Aids for identification of bivalve larvae of Virginia. Malacologia 11(1): 45-119.

Chuang, S. H. (1977). Larval development in *Discinisca* (inarticulate brachiopod). American Zoologist 17: 39-53.

Coe, W. R. (1926). The Pelagic Nemerteans. Mem. Mus. Comp. Zool. Harvard Coll. 49: 1-244.

Coe, W. R. (1954). The Bathypelagic Nemerteans of the Pacific Ocean. Bull. Scripps Inst. Oceanogr. 6: 225-286.

Coe, W. R. (1956). Pelagic Nemertea: keys to families and genera. Conseil international pour l'exploration de la mer Zooplankton sheet 64: 1-5.

Cook, H. L. and M. A. Murphy (1971). Early developmental stages of the brown shrimp, *Penaeus aztecus* Ives, reared in the laboratory. Fishery Bulletin 69(1): 223-239.

Costello, D. P. (1938). Notes on the breeding habits of the nudibranchs of Monterey Bay and vicinity. Journal of Morphology 63: 319-343.

Cupp, E. E. (1943). Marine Plankton Diatoms of the West Coast of North America. Berkeley, University of California Press.

Diaz, H. and J. D. Costlow (1972). Larval development of *Ocypode quadrata* (Brachyura: Crustacea) under laboratory conditions. Marine Biology 15: 120-131.

Dobkin, S. (1961). Early developmental stages of pink shrimp, *Penaeus duorarum* from Florida waters. Fishery Bulletin 190: 321-349.

Eckelbarger, K. J. (1976). Larval development and population aspects of the reef-building polychaete *Phragmatopoma lapidosa* from the east coast of Florida. Bulletin of Marine Science 26(2): 117-132.

Eckelbarger, K. J. (1977). Larval development of *Sabellaria floridensis* from Florida and *Phragmatopoma californica* from Southern California (Polychaeta: Sabellariidae), with a key to the sabellariid larvae of Florida and a review of development in the family. Bulletin of Marine Science 27(2): 241-255.

Emig, C. C. (1982). The biology of phoronida. *In* Advances in Marine Biology (J. H. S. Blaxter, F. S. Russell and M. Yonge, Eds.). San Francisco, Academic Press.

Emlet, R. B. and R. R. Strathmann (1994). Functional consequences of simple cilia in the mitraria of Oweniids (an anamolous larva of an anamolous polychaete) and comparisons with other larvae. *In* Reproduction and Development of Marine Invertebrates (W. H. J. Wilson, S. A. Stricker and G. L. Shinn, Eds.). Baltimore, The Johns Hopkins University Press.

References

Fenaux, R. (1967). Les Appendiculaires des mers d'Europe et du Bassin Méditerranéen. Faune de l'Europe et du Bassin Méditerranéen. Paris, Masson Ed.

Fensome, R. A., F. J. R. Taylor, et al. (1993). A Classification of Living and Fossil Dinoflagellates. Hanover, Sheridan Press.

Fraser, J. H. (1947a). Thaliacea - I. Conseil international pour l'exploration de la mer Zooplankton sheet 9: 1-4.

Fraser, J. H. (1947b). Thaliacea - II (Family: Doliolidae). Conseil international pour l'explaration de la mer Zooplankton sheet 10: 1-4.

Fraser, J. H. (1957). Chaetognatha. *In* Fiches d'Identification du Zooplancton (Conseil Permanent International pour l'Exploration de la Mer), No. 1 (revised).

Fretter, V. and M. C. Pilkington (1970). Prosobranchia: veliger larvae of taenioglossa and stenoglossa. Fiches D'identification du Zooplancton (129-132), 26 pp.

Fulton, J. (1968). A laboratory manual for the identification of British Colombia marine zooplankton. Fisheries Research Board of Canada, Pacific Oceanographic Group.

Gaines, G. and F. J. R. Taylor (1986). A mariculturist's guide to potentially harmful marine phytoplankton of the Pacific coast of North America. B.C. Ministry of Environment, Marine Resources Section, Fisheries Branch.

Galt, C. P. J. (1970). Population Composition and Annual Cycle of Larvacean Tunicates in Elliott Bay, Puget Sound. MS Thesis, University of Washington.

Gardner, G. A. and I. Szabo (1982). British Columbia pelagic marine Copepoda: An identification manual and annotated bibliography. Canadian Special Publication of Fisheries and Aquatic Sciences.

Gladfelter, W. B. (1972). Structure and Function of the Locomotory System of *Polyorchis montereyensis*. Sonderdruch Helgolander wiss. Meeresunters 23: 38-79.

Grave, C. (1944). The larva of *Styela* (Cynthia) *partita*: structure, activities and duration of life. Journal of Morphology 75(2): 173-188.

Green, J. C., K. Perch-Nielsen and P. Westbroek (1990). Phylum Prymnesiophyta. *In* Handbook of Protoctista (L. Margulis, J. O. Corliss, M. Melkonian and D. J. Chapman, Eds.). Boston, Jones & Bartlett.

Greene, R. W. (1968). The egg masses and veligers of Southern California Sacoglossan Opisthobranchs. The Veliger 11(2): 100-104.

References

Greer, D. L. (1962). Studies on the embryology of *Pycnopodia helianthoides* (Brandt) Stimpson. Pacific Science 16: 280-285.

Grimstone, A. V. and R. J. Skaer (1972). A Guidebook to Microscopical Methods. Cambridge, Cambridge University Press.

Hamner, W. M. (1974). Ghosts of the Gulf Stream Blue Water Plankton. National Geographic 146(4): 530-545.

Hannerz, L. (1961). Polychaeta: larvae. Families: Spionidae, Disomidae, Poecilochaetidae. Conseil international pour l'exploration de la mer Zooplankton sheet 91: 1-12.

Hardy, A. C. (1967). Great Waters: A Voyage of Natural History to Study Whales, Plankton, and the Waters of the Southern Ocean. New York, Harper & Row.

Hart, J. F. L. (1935). The larval development of British Columbia brachyura. Canadian Journal of Zoology 12(4): 411-432.

Hart, J. F. L. (1937). Larval and adult stages of British Columbia anomura. Canadian Journal of Research 15 Sec. D(10): 179-220.

Hart, J. F. L. (1960). The larval development of Bristish Columbia brachyura. Canadian Journal of Zoology 38: 539-546.

Hart, J. F. L. (1971). A key to planktonic larvae of families of decapod Crustacea of British Columbia. Syesis 4: 227-234.

Hatch, M. H. (1947). The Chelifera and Isopoda of Washington and adjacent regions. Univ. Wash. Publ. Biol. 10: 155-274.

Haynes, E. (1981). Early zoeal stages of *Lebbeus polaris, Eualus suckleyi, E. Fabricii, Spirontocaris arcuata, S. ochotensis,* and *Heptacarpus camtschaticus* (Crustacea, Decapoda, Caridea, Hippolytidae) and morphological characterization of zoeae of *Spirontocaris* and related genera. Fishery Bulletin 79(3): 421-440.

Haynes, E. B. (1984). Early zoeal stages of *Placetron wosnessenskii* and *Rhinolithodes wosnessenskii* (Decapoda, Anomura, Lithodidae) and review of Lithodid larvae of the Northern North Pacific Ocean. Fishery Bulletin 82(2): 315-324.

Haynes, E. B. (1985). Morphological development, identification, and biology of larvae of Pandalidae, Hippolytidae, and Crangonidae (Crustacea, Decapoda) of the Northern North Pacific Ocean. Fishery Bulletin 83(3): 253-288.

Humes, A. G. and A. G. Stock (1973). A Revision of the family Lichomolgidae Kossman, 1877, cyclopoid copepods mainly associated with marine invertebrates. Smithsonian Contrib. Zool. 127: 1-368.

References

Hurst, A. (1967). The egg masses and veligers of thirty Northeast Pacific opisthobranchs. The Veliger 9(3): 255-288.

Hustedt, F. (1927-1966). Die Keiselalgen. *In* Kryptogamen-Flora von Deutschland, Osterreich, und der Schweiz (L. Rabenhorst, Ed.).

Hyman, O. W. (1926). Studies of the larvae of crabs of the family xanthidae. Proceedings U.S. National Museum 67(3): 1-36.

Jägersten, G. (1963). On the Morphology and Behavior of Pelagosphaera Larvae (Sipunculoidea). Band 36, Zoologiska Bidrag Fran Uppsala, Uppsala.

Jahn, T. L. and F. F. Jahn (1949). How to Know the Protozoa. Dubuque, Wm. C. Brown Company.

Johns, D. M. and W. H. Lang (1977). Larval development of the spider crab, *Libinia emarginata* (Majidae). Fishery Bulletin 75(4): 831-841.

Knudsen, J. W. (1958). Life cycle studies of the brachyura of Western North America, I. General culture methods and the life cycle of *Lophopanopeus leucomanus leucomanusi* (Lockington). Bulletin, Southern California Academy of Sciences 57(1): 51-59.

Knudsen, J. W. (1959). Life cycle studies of the brachyura of Western North America, II. The life cycle of *Lophopanopeus bellus diegensis* Rathbun. Bull., Southern California Academy of Sciences 58(2): 57-64.

Knudsen, J. W. (1960). Life cycle studies of the brachyura of Western North America, IV. The life cycle of *Cycloxanthops novemdentatus* (Stimpson). Bull., Southern California Academy of Sciences 59(1): 1-8.

Korn, H. (1960). Introduction to the polychaete larvae. Universidade de Sao Paulo Instituto Oceanografico Catalogue of Marine Larvae (2): 3-88.

Kozloff, E. N. (1987). Marine Invertebrates of the Pacific Northwest. Seattle, University of Washington Press.

Kramp, P. L. (1965). The Hydromedusae of the Pacific and Indian Oceans, Section 1. Copenhagen, Andr. Fred Host & Son.

Kramp, P. L. (1968). The Hydromedusae of the Pacific and Indian Oceans, Sections II and III. Copenhagen, Andr. Fred Host & Son.

Kress, A. (1971). Uber die Entwicklung der Eikapselvolumina bei verschiedener Opisthobranchier-Arten (Mollusca, Gastropoda). Helgolander wissenschaftliche Meeresuntersuchungen 22: 326-349.

References

Kress, A. (1972). Veranderungen der eikapselvolumina wahrend der entwicklung verschiedener opisthobranchier-arten (Mollusca, Gastropoda). Marine Biology 16: 236-252.

Kress, A. (1975). Observations during embryonic development in the genus *Doto* (Gastropoda, Opisthobranchia). Journal of the Marine Biological Association of the U.K. 55: 691-701.

Lacalli, T. C. (1980). A guide to the marine flora and fauna of the Bay of Fundy: Polychaete larvae from Passamaquoddy Bay. Canadian Technical Report of Fisheries and Aquatic Sciences (940): 27 pp.

Lacalli, T. C. and J. E. West (1986). Ciliary band formation in the doliolaria larva of *Florometra* I. The development of normal epithelial pattern. J. of Embryology and Experimental Morphology 96: 303-323.

Lacalli, T. C. and J. E. West (1987). Ciliary band formation in the doliolaria larva of *Florometra* II. Development of anterior and posterior half-embryos and the role of the mesentoderm. Development 99: 273-284.

Lang, K. (1965). Copepoda harpacticoidea from the California Pacific coast. Kungl. Svenska Vetenskaps Akademiens Handlingar, Series 4 10(2): 1-566.

Lang, W. H. (1977). The Barnacle Larvae of North Inlet, South Carolina (Cirripedia: Thoracica). Ph.D. Thesis, University of South Carolina, Columbia.

Laubitz, D. (1970). Studies on the Caprellidae (Crustacea, Amphipoda) of the American North Pacific. Publ. Biol. Oceanogr., Mus. Nat. Sci. Canada 7. 89 pp.

Lea, H. E. (1955). The Chaetognaths of Western Canadian coastal waters. Journal of the Fisheries Research Board of Canada 12(4): 593-617.

Lebour, M. V. (1925). The Dinoflagellates of Northern Seas. Plymouth, The Marine Biological Association of the United Kingdom.

Lebour, M. V. (1927). Studies of the Plymouth brachyura. I. The rearing of crabs in captivity, with a description of the larval stages of *Inachus dorsettensis, Macropodia longirostris* and *Maia squinado*. Journal of the Marine Biological Association 14: 795-821.

Lebour, M. V. (1928). The larval stages of Plymouth Brachyura. Proceedings of the Zoological Society of London 96(1): 473-557.

Lee, J. J., H. H. Seymour and E.C. Bovee (1985). An Illustrated Guide to the Protozoa. Lawrence, Allen Press, Inc.

Levin, L. A. (1984). Multiple patterns of development in *Streblospio benedicti* Webster (spionidae) from three coasts of North America. Biological Bulletin 166: 494-508.

References

Lewis, C. A. (1975). Development of the gooseneck barnacle *Pollicipes polymerus* (Cirripedia: Lepadomorpha): fertilization through settlement. Marine Biology 32: 141-153.

Loosanoff, V. L., H. C. Davis and P.E. Chanley (1966). Dimensions and shapes of larvae of some marine bivalve mollusks. Malacologia 4(2): 351-435.

Lutz, R. et al. (1982). Preliminary observations on the usefulness of hinge structures for identification of bivalve larvae. Journal of Shellfish Research 2(1): 65-70.

Margulis, L., J. O. Corliss, M. Melkonian and D.J. Chapman (1990). Handbook of Protoctista. Boston, Jones & Bartlett.

Marshall, S. M. (1969). Protozoa. Conseil international pour l'exploration de la mer Zooplankton sheets 117-127.

Masterman, A. T. (1902). The early development of *Cribrella oculata* (Forbes), with remarks on echinoderm development. Trans. Roy. Soc. Edinburgh 40: 373-418.

Matarese, A. C., A. W. Kendall Jr., D.M. Blood and B.M. Vinter (1989). Laboratory Guide to Early Life History Stages of Northeast Pacific Fishes. NOAA Technical Report NMSF 80.

McEuen, F. S. and F. S. Chia (1985). Larval development of a molpadiid holothuroid, *Molpadia intermedia* (Ludwig, 1894) (Echinodermata). Canadian Journal of Zoology 63: 2553-2559.

Miller, K. M. and J. Roughgarden (1994). Descriptions of the larvae of *Tetraclita rubescens* and *Megabalanus californicus* with a comparison of the common barnacle larvae of the central California coast. Journal of Crustacean Biology 14(3): 579-600.

Mills, C. E. (1987). Hydromedusae. *In* Marine Invertebrates of the Pacific Northwest (E. N. Kozloff, Ed.). Seattle, University of Washington Press.

Mladenov, P. V. and F. S. Chia (1983). Development, settling behavior, metamorphosis and pentacrinoid feeding and growth of the feather star *Florometra serratissima*. Marine Biology 73: 309-323.

Montagnes, D. J. S. and F. J. R. Taylor (1994). The salient features of five marine ciliates in the class Spirotrichea (Oligotrichia), with notes on their culturing and behaviour. Journal of Eukaryotic Microbiology 41(6): 569-586.

Moore, A. R. (1943). On the embryonic development of the sea urchin *Allocentrotus fragilis*. Biological Bulletin 117: 492-496.

Mortensen, T. (1921). Studies of the Development and Larval Forms of Echinoderms. Copenhagen, G.E.C. Gad.

References

Mortensen, T. (1927). Handbook of the Echinoderms of the British Isles. London, Oxford University Press.

Naumov, D. V. (1969). Hydroids and Hydromedusae of the USSR. English translation 1969 by J. Salkind. Jerusalem, IPST Press.

Newell, G. E. and R. C. Newell (1967). Marine Plankton: A Practical Guide. London, Hutchinson Educational, Ltd.

Nielsen, C. (1987). Structure and function of metazoan ciliary bands and their phylogenetic significance. Acta Zoologica 68(4): 205-262.

Nielsen, C. (1990). The development of the brachiopod *Crania (Neocrania) anomala* (O.F. Müller) and its phylogenetic significance. Acta Zoologica, Stockh. 72: 7-28.

Ockelmann, K. W. (1962). Developmental types in marine bivalves and their distribution along the Atlantic coast of Europe. *In* Proceeding of the 1st European Malacological Congress (L. R. Cox and J. F. Peake, Eds.). London, Conchological Society Great Britain & Ireland and Malacological Society London. 25-35.

O'Donoghue, C. H. and E. O'Donoghue (1922). Notes on the nudibranchiate mollusca from the Vancouver Island region. II. The spawn of certain species. Transactions of the Royal Canadian Institute 14: 131-143.

Page, L. R. (1994). The ancestral gastropod larval form is best approximated by hatching-stage opisthobranch larvae: evidence from comparative developmental studies. *In* Reproduction and Development of Marine Invertebrates (W. H. J. Wilson, S. A. Stricker and G. L. Shinn, Eds.). Baltimore and London, The Johns Hopkins University Press.

Pearse, J. S. (1979). Polyplacophora. *In* Reproduction of Marine Invertebrates (A. C. Giese and J. S. Pearse, Eds.). San Francisco, Academic Press.

Pike, R. B. and D. I. Williamson (1960). The larvae of *Spirontocaris* and related genera (Decapoda, Hippolytidae). Crustaceana 2: 13-208.

Poole, R. L. (1966). A description of laboratory-reared zoeae of *Cancer magister* Dana, and megalopae taken under natural conditions (decapoda brachyura). Crustaceana 11: 83-97.

Rees, C. B. (1950). The identification and classification of lamellibranch larvae. Hull Bulletins of Marine Ecology 3(19): 73-104.

Reeve, M. R. (1981). Large cod-end reservoirs as an aid to the live collection of delicate zooplankton. Limnology and Oceanography 26(3): 577-580.

Rice, A. L. (1980). Crab zoeal morphology and its bearing on the classification of the Brachyura. Trans. Zool. Soc. Lond. 35: 271-424.

References

Rice, M. E. (1967). A comparative study of the development of *Phascolosoma agassizii, Golfingia pugettensis,* and *Themiste peroides* with a discussion of developmental patterns in the sipuncula. Ophelia 4: 143-171.

Rice, M. E. (1975a). Observations on the development of six species of Caribbean Sipuncula with a review of development in the phylum. *In* Proceedings of the International Symposium on the Biology of the Sipuncula and Echiura (M. E. Rice and M. Todorovic, Eds.). Belgrade, Naucno Delo Press.

Rice, M. E. (1975b). Sipuncula. *In* Reproduction of Marine Invertebrates (A. C. Giese and J. S. Pearse, Eds.). San Francisco, Academic Press.

Rice, M. E. (1976). Larval development and metamorphosis in Sipuncula. American Zoologist 16: 563-571.

Richardson, H. (1905). A monograph of the Isopods of North America. Bull. 54, U.S. Nat. Mus.

Richter, G. and G. Thorson (1975). Pelagische prosobranchier-larven des golfes von neapel. Ophelia 13: 109-185.

Robertson, R. (1974). Marine prosobranch gastropods: larval studies and systematics. Thalassia Jugoslavica 10(1/2): 213-238.

Roesijadi, G. (1976). Descriptions of the prezoeae of *Cancer magister* Dana and *Cancer productus* Randall and the larval stages of *Cancer antennarius* Stimpson (decapoda, brachyura). Crustaceana 31(3): 275-295.

Round, F. E. and R. M. Crawford (1990). Phylum Bacillariophyta. *In* Handbook of Protoctista (L. Margulis, J. O. Corliss, M. Melkonian and D. J. Chapman, Eds.). Boston, Jones & Bartlett.

Ryland, J. S. (1974). Behaviour, settlement and metamorphosis of bryozoan larvae: a review. Thalassia Jugoslavica 10(1/2): 239-262.

Schlotterbeck, R. E. (1976). The larval development of the lined shore crab, *Pachygrapsus crassipes* Randall, 1840 (decapoda, brachyura, grapsidae) reared in the laboratory. Crustaceana 30(2): 184-200.

Schultz, G. A. (1969). How to Know the Marine Isopod Crustaceans. Dubuque, Wm. C. Brown.

[De] Schweinitz, E. H. and R. A. Lutz (1976). Larval development of the Northern horse mussel, *Modiolus modiolus* (L.), including a comparison with the larvae of *Mytilus edulis* L. as an aid in planktonic identification. Biological Bulletin 150: 348-360.

Sherr, E. B. and B. F. Sherr (1991). Planktonic Microbes: tiny cells at the base of the ocean's food webs. Trends in Ecology and Evolution 6(2): 50-54.

References

Smiley, S. (1986a). Metamorphosis of *Stichopus californicus* (Echinodermata: Holothuroidea) and its phylogenetic implications. Biological Bulletin 171: 611-631.

Smiley, S. T. (1986b). *Stichopus californicus* (Echinodermata: Holothuroidea) Oocyte Maturation Hormone, Metamorphosis, and Phylogenetic Relationships. Ph.D. Thesis, U. of Washington.

Sournia, A. (1978). Phytoplankton Manual. Paris, UNESCO Press.

Stebbing, T. R. R. (1906). Amphipoda, 1: Gammaridea. Das Tierreich, pt. 21. 806 pp.

Steedman, H. F. (1976). Zooplankton Fixation and Preservation. Paris, UNESCO Press.

Steidinger, K. A. and K. Tangen (*In Press*). Dinoflagellates. *In* Identifying Marine Diatoms and Dinoflagellates (C. Tomas, Ed.). San Diego, Academic Press.

Strathmann, M. (1987). Reproduction and Development of Marine Invertebrates of the Northern Pacific Coast. Seattle, University of Washington Press.

Strathmann, R. R. (1971). The feeding behavior of planktotrophic echinoderm larvae: mechanisms, regulation, and rates of suspension-feeding. The Journal of Experimental Marine Biology and Ecology 6: 109-160.

Strathmann, R. R. (1979). Echinoid larvae from the Northeast Pacific (with a key and comment on an unusual type of planktotrophic development). Canadian Journal of Zoology 57: 610-616.

Sullivan, C. (1948). Bivalve larvae of Malpeque Bay, P.E.I. Bulletin, Fisheries Research Board of Canada 77: 1-36.

Svane, I. (1982). Possible ascidian counterpart to the vertebrate Saccus Vasculosus with reference to *Pyura tessellata* (Forbes) and *Boltenia echinata* (L.). Acta Zoologica 63(2): 85-89.

Taylor, J. (1975). Planktonic Prosobranch Veligers of Kaneohe Bay. Ph.D. Thesis, U. of Hawaii.

Taylor, F. J. R., W. A. S. Sarjeant, R.J. Fensome and G.L. Williams (1987). Standardisation of nomenclature in flagellate groups treated by both the botanical and zoological codes of nomenclature. Systematic Zoology 36(1): 79-85.

Thiriot-Quievreux, C. (1980). Identification of some planktonic prosobranch larvae present off Beaufort, North Carolina. The Veliger 23(1): 1-9.

Thiriot-Quievreux, C. (1983). Summer meroplanktonic prosobranch larvae occurring off Beaufort, North Carolina. Estuaries 6(4): 387-398.

Thiriot-Quievreux, C. and R. Scheltema (1982). Planktonic larvae of new england gastropods. v. *Bittium alternatum, Triphora nigrocincta, Cerithiopsis emersoni, Lunatia heros* and *Crepidula plana*. Malacologia 23(1): 37-46.

References

Thorson, G. (1934). On the reproduction and larval stages of the brittle-stars *Ophiocten sericeum* (Forbes) and *Ophiura robusta* Ayres in East Greenland. Meddelelser om Grønland Udgivne af Kommissionen for Videnskabelige Undersøgelser I Grønland 100(4): 1-21.

Thorson, G. (1940). Studies on the egg-masses and larval development of Gastropoda from the Iranian Gulf. Danish Scientific Investigations in Iran 2: 159-238.

Trask, T. (1970). A description of laboratory-reared larvae of *Cancer productus* Randall (decapoda, brachyura) and a comparison to larvae of *Cancer magister* Dana. Crustaceana 18: 133-146.

Uchida, T. (1927). Studies on Japanese Hydromedusae I. Anthomedusae. Journal of the Faculty Sci. University of Tokyo Section IV. Zoology 1(3): 145-241.

Vannucci, M. (1961). Introduction to the echinodermata - echinoidea larvae. Universidade de Sao Paulo Instituto Oceanografico: Catalogue of Marine Larvae (3): 1-13.

Vinyard, W. C. (1975). Key to the Genera of Marine Plankton Diatoms of the Pacific Coast of North America. Eureka, CA, Mad River Press.

Watanabe, J. and L. Cox (1975). Spawning behavior and larval development in *Mopalia lignosa* and *Mopalia muscosa* (Mollusca: Polyplacophora) in Central California. Veliger 18, supp.: 18-27.

Williamson, D. I. (1982). Larval morphology and diversity. *In* The Biology of Crustacea: Embryology, Morphology and Genetics (L. Abele, Ed.). San Francisco, Academic Press. 43-111.

Wilson, D. P. (1932). On the Mitraria Larva of *Owenia fusiformis*. Philosophical Transactions of the Royal Society of London, Series B v. 221.

Wilson, D. P. (1982). The larval development of three species of *Magelona* (Polychaeta) from the localities near Plymouth. Journal of the Marine Biological Association of the U.K. 62: 385-401.

Wimpenny, R. S. (1966). The Plankton of the Sea. New York, Elsevier.

Wood, E. J. F. (1965). Marine Microbial Ecology. New York, Chapman & Hall.

Yamashita, M. (1985). Embryonic development of the brittle-star *Amphipholis kochii* in laboratory culture. Biological Bull. 169: 131-142.

Zimmer, R. (1964). Reproductive Biology and Development of Phoronida. Ph.D. Thesis, U. of Washington.

Zimmer, R. and R. Woollacott (1977). Structure and classification of gymnolaemate larvae. *In* Biology of Bryozoans (Woollacott and Zimmer, Eds.). San Francisco, Academic Press. 57-89.

Index

Abdominal appendages, 120, 132, 139, 145
Abietinaria, **76**, 78
Acartia, **123**
Achelia, 111, **112**
Achnanthes, **42**
Acorn Worm, 183
Actinopoda, **48**
Actinotroch, 169, **170**, 197
Actinula, 58, **64**, 65, 69, 75, 197
Adductor Muscle, 161, **164**, 197
Adult, 10, 55, 56, 58, 61, 65, 83, 85, 89, 97, 104, 109, 111, 113, 115, 129, 132, 139, 141, 145, 153, 155, 161, 163, 167, 171, 173, 175, 176, 180, 183, 185, 187, 192, 197, 198, 199, 200, 201, 202
Aegina, 69, **70**
Aglantha, 69
Aglaophenia, **77**, 78
Algae, 1, 14, 25, 26, 29, 49, 104, 111, 136, 155, 198
Amphiblastula, 63, 197
Amphinema, **74**
Amphipholis, 183
Amphipoda, 129, **135**, 136, **137-138**
Ampithoe, **137**
Ampullae, 187, 197
Ancestrula, 169, 197
Anchovy, 192, **193**
Anemone, 57, 63, 65, 111, 202
Annelida, 97, **100-103**, 104, **105-108**, 109, 200, 201
Anomura, 141, 145, **148**, **150**
Antennae, 4, 120
Anthomedusa, 67, 68, 69, **72-74**, 75, 199
Anthopleura, 57
Anthozoa, 63
Aphotic Zone, 3, 197
Apical Tuft, 85, 97, 99, 169, 197
Appendicularia, **189**
Arachnida, 111, **112**
Arachnoidiscus, **30**
Arrow Worm, 173
Arthropoda, 59, 111, **112**, **114**, **116-118**, **122-128**, **130-131**, **133-135**, **136-138**, **140**, **142-144**, **146-151**, 198, 200
Ascidiacea, 185, **186**, 187, **188**, 197, 202
Asexual Reproduction, 49, 57, 63, 69, 75, 78, 104, 167, 169, 187, 198, 199, 202
Asterionella, **39**
Asteroidea, 175, 176, **177-178**, 197
Asterolampra, **35**
Atherinidae, **194**
Atoke, 104
Attenuation, 197
Aurelia, **64**, 65, **66**, 136
Auricularia, **179**
Autolytus, 99, **101**
Autotrophic, 3, 26, 49, 197, 201
Auxospore, 29, 197

Bacillaria, **42**
Bacillariophyceae, **42**, 43
Bacteriastrum, **36**
Balanus, 115
Barnacles, 57, 115, 198
Baseodiscus, 89
Bean Clam, 115
Bell, 56, 67, 68, 69, 78, 197, 200, 202
Benthic, 1, 10, 46, 55, 57, 60, 65, 89, 104, 111, 113, 120, 129, 132, 136, 155, 161, 167, 171, 173, 175, 180, 183, 185, 192, 197
Benthic Adult, 10, 55, 57, 104, 161, 167, 171, 173, 175, 180, 183
Beröe, **84**
Biddulphia, **33**
Bioluminescence, 119
Bipinnaria, 59, 176, **177-178**, 197
Biramous, 113, 129, 132, 197, 198
Bivalvia, 115, 153, 161, **164**, 197, 199, 201, 202
Blastopore, 61, 175, 197, 198, 201
Blastula, 55, 175, 197, 198, 199
Blepharipoda, 145, **148**
Bloom, 10, 27, 49, 197
Bolinopsis, **84**, 85
Bopyrus, 132
Bougainvillia, **74**, 75
Bowerbankia, **168**, 169
Brachiolaria, 59, 176, **178**, 197
Brachionus, **94**
Brachiopoda, 61, 167, 171, **172**, 199, 200
Brachyura, 141, 145, **148-151**
Bract, 78, **80-81**, 197
Branchial chamber, 185
Bristles, 202
Brittle Star, **182**, 183, 197, 200
Brood, 57, 99, 119, 129, 132, 136, 183, 197, 198
Bryozoa, 57, 58, 167, **168**, 169, 197, 198, 200, 202
Budding, 69, 75, 78, 104, 169, 187, 197, 199, 202
Bugula, **168**, 169

Calanoid Copepod, 120, 121, **123-124**
Calanus, **124**
Calcareous, 46, 49, 109, 199
Caligus, **127**
Calycophora, 198, 202
Calyptoblastae, 75, 197
Cancer, 145
Capitellidae, 104, **107**
Caprella, **138**
Caprellidae, 136, **138**
Carapace, 57, 109, 115, 119, 129, 132, 139, 141, 145
Carcinonemertes, 89
Caridea, 141, **142-144**
Carnivore, 3, 56, 104, 173
Centric Diatoms, 43
Centropages, **124**
Cephalization, 113, 197
Cephalopoda, 58, 153, 163, **165**, 166
Cephalothorax, 197

215

Index

Ceratium, **51**
Chaetae, 99, 109, 197
Chaetoceros, 14, **36-38**, 43
Chaetognatha, 55, 61, 89, **172**, 173, 199
Chaetopteridae, 104, **105**
Chaetopterus, 104, **105**
Chain, 3, 16, 29, 78, 109, 169, 202
Chelae, 129, 132, 139, 141
Chelate, 141
Chelicerata, 111
Chitin, 56, 75, 109
Chiton, 153, **154**
Chloroplast, 29, 49, 198
Chondrophora, 78, **79**
Chordata, 55, 59, 61, 166, 183, 185, **186**, 187, **188-189**, 190, **191**, 192, **193-196**, 197, 200, 202
Chorion, 153, **154**, 192, 198
Chromatophore, 141, 166, 192, 198
Chrysaora, 65, **66**
Chrysopetalidae, **106**
Cilia, 4, 43, 56, 63, 83, 85, 93, 97, 109, 169, 175, 183, 197, 198, 199
Ciliophora, 25, 28, 43, **44-45**, 46, 197, 200
Cirratulidae, **105**
Cirripedia, 57, 115, **116**, 198
Cladocera, **118**, 119
Cladonema, 69, **73**
Clams, 49, 115, 153, 161, 197, 201
Cleavage, 59, 61, 155, 166, 176, 201, 202
Climacocodon, **73**
Climacodium, **34**
Clione, 158, **160**
Clytia, 67, 68
Cnidae, 63
Cnidaria, 56, 58, 63, **64**, 65, **66**, **70-74**, **76-77**, **79-81**, 111, 119, 197, 198, 199, 200, 201, 202
Coccolithophore, 28
Coeloblastula, 63, 198
Collection, 5, 10, 11, 25, 69, 155, 198
Colloblasts, 83, 198
Comb Jelly, 12, 56, 58, 83, **84**, 89, 136, 198, 200
Commensal, 111
Compound eye, 113, 119, 129, 141
Conchoecia, **117**, 119
Consumers, 26, 27
Continuous plankton recorder, 8
Copepoda, 12, 15, 27, 55, 57, 113, 119, 120, 121, **122-128**
Coral, 63, 65
Corbula, 78, 198
Corethron, **35**
Cormidium, 78, **82**, 197, 198, 202
Corona, 93, 198
Coronate Larva, 58, **168**, 169, 198
Corraline algae, 155
Coscinodiscus, 29, **30**
Coscinodiscophyceae, **30-38**
Crabs, 113, 129, 132, 139, 145
Crania, 173
Crossota, 69
Crowding, 11, 13

Crustacea, 1, 58, 109, 111, 113, 115, 119, 129, 132, 199, 200, 201, 202
Cryptomonas, **50**
Cryptoniscid Larva, 132, **134**, 198
Cryptopontius, **125**
Ctenes, 83, 198
Ctenophora, 12, 56, 58, 83, **84**, 85, 89, 136, 198
Cubozoa, 63
Culture, 5, 13, 14, 61, 109
Cultures, 5, 12, 13, 14, 61, 104, 109
Cumacea, 129, 132, **133**
Currents, 3, 27, 111, 167, 200, 201, 202
Cyanea, 65, 136
Cyclopoida, 120, 121, **125**
Cyclops, 119
Cydippid Larva, 83, 85
Cylindrotheca, **42**
Cyphonautes, 58, 167, **168**, 169, 198
Cypridina, **117, 118**
Cypris Larva, 59, 115, **116**, 198

Daphnia, **118**
Decapoda, 139, 145, 200, 201
Decomposition, 27
Demersal, 120, 192, 198
Dendraster, **181**
Dendrobranchiata, 141
Density, 9, 69, 192
Deuterostome, 61, 166, 173, 183, 198, **Plts. 97-112**
Development, 55, 58, 60, 83, 97, 104, 109, 115, 132, 139, 141, 145, 153, 155, 161, 163, 166, 169, 171, 173, 175, 176, 180, 181, 183, 192, 198, 199
Diarthrodes, 121, **126**
Diatoms, 1, 3, 4, 14, 15, 16, 26, 27, 28, 29, **30-42**, 43, 53, 104, 119, 197, 198, 199, 201
Dictyocha, **50**
Dinoflagellata, 4, 26, 28, 49, **51-52**, 53
Dioecious, 97, 175, 198
Diphyes, **79**
Diplosoma, 187
Direct Development, 58, 132, 163, 198
Discinisca, 173
Discorbis, **47**
Ditylum, **32**
Diversity, 10, 15, 27, 29, 97, 113, 185
Doliolaria, 59, 176, **179**, 180, 198
Doliolum, 190, **191**
Dunaliella, 14, **50**

Echinodermata, 56, 59, 61, 166, 175, 176, **177-179**, 180, **182**, 183, 197, 198, 200, 201
Echinoidea, 175, 180, **181**, 183, 198
Echinopluteus, 59, 180, **181**, 183, 198
Echinospira Larva, 158, **159**, 198
Echiura, 93, **96**, 97
Ectoprocta, 61, 167
Egg case, 57, 155, **156**, 161, 163
Eggs, 1, 4, 27, 55, 57, 59, 65, 75, 78, 85, 99, 104, 119, 120, 129, 132, 139, 145, 153, 155, 158,

Index

161, 163, 167, 173, 175, **177**, 192, 198, 199, 200
Embryos, 4, 10, 57, 58, 139, 155, 166, 169, 176, 201
Emerita, 145, **148**
Emiliania, **50**
Endoderm, 166, 175, 198
Energy, 3, 26, 27, 197, 198, 199, 201, 202
Engraulis, **193**
Ephyra, **64**, 65, 198, 202
Epilabidocera, **124**
Epitheca, 29, 198
Epitoke, 99, **101**, 198
Erpochaete, **108**
Eucampia, **31**
Eucarida, 129, 139
Eucoel, 166, 198
Eudistylia, 97
Eudoxid, 78, 198
Eukrohnia, **172**, 173
Eunicidae, 99
Euphausiacea, 139, **140**
Euphotic, 3, 27, 199, 202
Euphysa, **74**
Eupolymnia, 104, **107**
Eutonina, 68, **71**
Evadne, **118**, 119
Excurrent siphon, 185
Exogone, 99, **101**
Exoskeleton, 75, 201
Exumbrella, 67, 69, 75, 199

Fairy Shrimp, 141
Fecal Pellet, 11
Feeding, 1, 4, 14, 27, 49, 56, 93, 97, 109, 115, 167, 169, 171, 175, 176, 180, 199, 201, 202
Fertilization, 57, 61, 85, 175, 199
Fertilization membrane, 61, 85, 175
Fish Egg, 55, **193-194**
Fish Larvae, **193-196**
Fission, 57
Flagella (um), 49, 199
Flagellates, **50**
Flatworm, 89, 200
Florometra, 180
Flowmeter, 9
Food Web, 1, 3, 27, 198, 200
Foot, 93, 161, 201
Foraminifera, 16, 28, 46, **47**, 49, 199
Foraminiferan Ooze, 46, 49
Fragilaria, **41**
Fragillariophyceae, **39-41**
Fragment, 26, 75, 78, 83, 167
Franklin Point, 183
Fritillaria, 187, **189**
Frustule, 29, 198, 199
Fusiform, 93, 199

Gametes, 55, 57, 65, 99, 173, 175, 183

Gammaridea, **135**, 136, **137**
Gastropoda, 57, 61, 153, 155, **156-157**, 158, **159-160**, 161, 198, 202
Gastropteron, 158
Gastrovascular Cavity, 67, 69, 78, 199, 201
Gastrula, 1, 55, 61, 166, 169, 175, **177**, **179**, 199
Ghost Shrimp, 141
Gills, 4, 132, 136, 139, 141, 161, 185, 192, 197
Girdle, 29
Globigerina, **47**
Globular, 68
Glossobalanus, 183
Glottidia, 173
Glycinde, **103**
Gnorisphaeroma, 132, **134**
Golfingia, 93, 97
Gonads, 68, 99
Gonangium, 75, 200
Goniadidae, **103**
Gonophore, 78, **80**, 199
Gonotheca, 75, 199
Gonothecae, 75, 199
Gonyaulax, 49, **52**
Grammatophora, **41**
Grazers, 3
Gymnoblast, 75, 199
Gymnodinium, **52**
Gymnosomate, 158

H_2O, 2, 5, 6, 7, 8, 9, 10, 11, 12, 13, 14, 27, 43, 46, 56, 57, 63, 65, 83, 89, 93, 111, 119, 121, 136, 155, 158, 161, 167, 185, 192, 201, 202
Halitholus, **73**
Harpacticoida, 120, 121, **126**
Harpacticus, **126**
Head, 93, 113, 120, 129, 197, 198
Heart, 129, 161, 192
Helgoland Net, 7
Hemichordata, 59, 61, 183, **186**, 202
Hensen Net, 7, 8
Herbivores, 3, 56, 119
Hermaphroditic, 97, 173, 199, 200
Hermit Crabs, 141, 145
Heterotrophic, 3, 4, 28, 199
Hinge, 163, 199
Hippolytidae, 141
Hirudinea, 97, **108**, 109
Holoplankton, 2, 55, 57, 199
Holothuroidea, 176, **179**, 198, 201
Hoplocarida, **133**, 139
Hormiphora, **84**, 85
Hyalodiscus, **31**, **35**
Hybocodon, 69, **72**
Hydra, 119
Hydractinia, **77**
Hydroidea, 57, 67, **70-74**, 75, **76**, **77**, 78, 111, 136, 167, 199, 200, 201
Hydromedusa, 56, 65, 67, 68, 199
Hydromedusae, 56, 65, 67, 68, 199
Hydrothecae, 75, 199
Hydrozoa, 57, 63, 65, 67, 75, 78, 167, 197, 199, 200

217

Index

Hypothecae, 29, 199

Idotea, 132, **134**
Indirect Development, 58, 199
Insecta, 111, 139
Inter-radial, 69, 166, 199
Invagination, 175, 197
Invertebrate Embryology, 55
Invertebrates, 1, 10, 14, 27, 55, 56, 57, 58, 59, 60, 104, 153, 171, 175
Isochrysis, 14, **50**
Isogametes, 57, 199
Isopoda, 109, 120, 129, 132, **134-135**, 136, 198, 200
Isthmia, **31**

Jelly, 83, 136, 200
Juvenile, 85, 104, 120, 145, 158. 161, 163, 166, 171, 176, 181, 183, 187

Kelp Crabs, 145
Kitahara Net, 8
Krill, 139

Lancelets, 185
Larvacea, 56, 187, **189**
Larvae, 1, 4, 10, 14, 27, 55, 56, 58, 63, 65, 75, 78, 83, 85, 89, 93, 97, 99, 104, 111, 113, 115, 120, 132, 139, 141, 145, 153, 155, 158, 161, 167, 169, 171, 175, 176, 180, 181, 183, 185, 187, 192, 197, 198, 200, 201, 202
Laterally Compressed, 120, 136
Lauderia, **33**
Lecithotrophic, 180, 199
Leech, 97, **108**, 109
Lensia, **80**
Leptocylindricus, **36**
Leptomedusae, 67, 68, **71**, 75, 197
Leuckartia, **74**
Libinia, 153
Licmorpha, **40**
Light, 3, 10, 11, 27, 173, 197, 199, 201
Limacina, 158, **160**
Limnomedusae, 67, 68
Limpets, 155, 161
Lingula, 173, 199
Lingula Larva, 171, **172**, 199
Liriope, **70**
Littorina, 155, **156**
Lobate Larva, **172**
Lobsters, 129, 139, 141
Loligo, 163, **165**
Lophophorate, 61, 167, **168, 170**, 171, **172**, 173, 200
Lophophore, 167, 171, 200

Lorica, **45**
Lumbrineridae, **108**

Macroplankton, 2
Macropsis, **131**
Magelona, **105**
Magelonidae, **105**
Malacostraca, 129, 139
Manubrium, 67, 68, 75, 200
Margin, 67, 68, 69, 200
Marine Food Web, 3
Marutoku, 8
Maxillopoda, 115, 119
Medusae, 12, 56, 57, 65, 67, 68, 69, 75, 78, 89, 197, 198, 199, 200, 201, 202
Megalopa, 59, 145, **150, 151**, 200
Megaloplankton, 2, 3
Melosira, 29, **31**
Membranipora, 167, 169
Meridonal, 83, 200
Meroplankton, 2, 3, 55, 57, 58, 200
Mesh Sizes, 2, 3, 5, 6, 8, 9, 10, 192
Mesoderm, 166, 200
Mesoplankton, 2
Messenger, 6
Metacaprella, **138**
Metamorphosis, 55, 56, 89, 97, 104, 113, 145, 153, 158, 161, 163, 169, 171, 180, 183, 187, 199, 200, 201
Metanauplius, 59, 120, **122**
Metandrocarpa, 187, **188**
Metatrochophore, 99, **101-103, 105-108**, 200
Microbial food web, 3, 27
Microniscid larva, 200
Microplankton, 2, 3
Microstoma, 89, **91**
Mitraria Larva, 104, 111, 200
Mollusca, 59, 61, 153, **154, 156, 157, 159, 160, 162, 164, 165,** 197, 201, 202
Monochrysis, 14
Monoecious, 85, 173, 199, 200
Monterey Bay, 27
Moss animals, 57, 58, 167, 169, 197, 198, 200, 202
Motile, 4, 25
Mouth, 5, 6, 7, 8, 9, 56, 67, 69, 75, 83, 97, 129, 167, 173, 199, 200, 201
Mucous, 43, 56, 171
Muggiaea, **79**
Müller's Larva, **90**
Munna, 132, **134**
Mussels, 49, 161
Myctophidae, **195**
Myotomes, 192, 200
Mysid, 109, 129, **130-131**, 200
Mysidacea, 109, 129, 200
Mytilus, 163, **164**

Nannoplankton, 2
Nanomia, **80**

Index

Narcomedusae, 67, 68, 69
Nauplius Larva, 59, 113, **114**, 115, **116**, 120, **122**, **140**, 200
Navicula, **42**
Neanthes, 97
Nectochaete, 99, **103**, 200
Nectophore, 78, **80**, 200
Nekton, 2
Nematocysts, 56, 67, 68, 69, 200
Nematoda, 93, **94**
Nemertea, 57, 58, 85, **86-88**, 201
Neomysis, **130**
Nephtys, 97, **103**
Nereidae, 99, **103**, 200
Nereis, 97
Neritic, 2, 200
Neurotroch, 99, 200
Nipponnemertes, 89
Nitrates, 26
Noctiluca, **51**
North Pacific, 75, 93, 109, 115, 119, 139, 173, 183, 187
Northern Pacific, 75, 85, 93, 109, 115, 119, 139, 173, 183, 187
notochord, 200
Nutrients, 3, 26, 27, 200, 201
Nymphon, **112**

Obelia, 68, **70**, 75, **76**
Observation, 5, 11, 12, 13, 109, 136, 169
Ocelli, 68, 69
Octocoral, 63
Octopus, 58, 163, **165**, 166
Odontella, **33**
Oikopleura, 56, 187, **189**
Oithona, 120, **125**
Oligochaete, 97
Oncaea, **121**
Ophiodromus, **108**
Ophiopluteus, 59, **182**, 183, 200
Ophiuroidea, **182**, 183, 197, 200
Ophiothrix, **182**
Ostracoda, 115, **117, 118**
Owenia, 104, **107**, 111, 200
Oxygen, 11, 13, 26, 161

Pachygrapsus, 153
Paddle system, 14
Paguridae, 145, **146, 147, 150**
Pagurus, **146, 150**
Palps, 109
Panulirus, 145
Papillae, 200
Paracalanus, **123**
Parafavella, **44**
Paralichthyidae, **193**
Paranemertes, 89
Parapodia, 56, 99, 109
Parasite, 89

Parasitic, 89, 109, 111, 120, 121, 132, 173, 198, 200
Parenchymula, 201
Pectinariidae, **105**
Pediveliger, 59, 163, **164**, 201
Pelagia, 65
Pelagic, 2, 65, 75, 89, 99, 111, 113, 121, 136, 155, 158, 201
Pelagosphaera Larva, 59, 93, **95**, 201
Pelecypoda, 161
Penaeidae, 141
Penaeus, 141
Pennate, 43
Pentacula, 59, 176, **179**, 201
Peridinium, **52**
Periopods, 129, 132, 139, 201
Peristaltic, 129
Phaeodactylum, 14
Phascolosoma, 93, 97
Phialella, 68, **71**
Phialidium, 67, 68, **71**
Phoronida, 58, 61, 167, 169, **170**, 171, 197, 200
Phoronis, 171
Phoronopsis, 171
Phosphates, 26
Photochemical, 26
Photosynthesis, 26, 29, 198, 199
Photosynthetic, 3, 49
Phronima, 136
Phyllaplysia, 161, **162**
Phyllodocidae, 104, **106**
Phylogenetics, 25, 61, 200, 201
Physonectae, 78
Phytoplankton, 2, 4, 8, 10, 11, 12, 14, 26, 27, 49, 53, 197, 200, 201
Picoplankton, 2, 3
Pigment, 12, 49, 141, 155, 166, 198
Pilidium, 58, 85, **86, 87**, 201
Pinnixa, 153
Plankton Net, 2, 3, 5, 7, 8, 9, 26, 65, 89, 141, 190, 198, 199, 200
Planktoniella, **35**
Planktotrophic, 27, 56, 104, 176, 180, 201
Planula larva, 58, 63, **64**, 65, 78, 83, 198, 201
Platyhelminthes, 58, 89, **90, 91**
Platynereis, **103**
Pleopods, 129, 139, 145, 201
Pleurobrachia, 83, **84**
Pleuronectidae, **195**
Pleuronichthys, 192, **194**
Pleurosigma, **42**
Plumularia, **76**
Pluteus, 180, 181, 183, 198, 200, 201
Podocoryne, **71**
Podon, **118**, 119
Pollicipes, 115
Pollution, 11
Polychaete, 15, 56, 57, 97, **100-103**, 104, **105-108**, 109, 197, 198, 200, 201, 202
Polycladida, 89, **91**
Polygordius, 104, **107**
Polynoidae, **106**
Polyorchis, 69, **73**
Polyp, 57, 65, 67, 69, 75, **76-77**, 78, 199, 201, 202

219

Index

Porcellanidae, 145, **148, 150**
Porifera, 63, 197, 201
Predation, 104, 173
Predators, 3, 11, 109
Pre-zoea, 145, **149**, 201
Primary productivity, 4, 26, 27, 28, 198
Primitive gut, 175
Proboscidactyla, 68, **70**
Producers, 3, 26, 27
Productivity, 26
Protista, 1-4, 6, 8, 12, 25, 26, 28, 199, 201
Protoperidinium, **52**
Protoplast, 29
Protostomes, 61, 201
Prototroch, 99, 153, 201
Pseudocalanus, **124**
Pseudo-nitzschia, **42**
Psolus, **179**, 180
Pteropoda, 158, **160**
Purple Sailor, 78
Pycnogonida, 111, **112**
Pycnogonum, 111, **112**
Pycnopodia, 176
Pygidium, 57, 104, 109, 201
Pyura, 187

Radial, 61, 67, 68, 69, 75, 166, 176, 201
Radial canal, 67, 68, 69, 75
Radial cleavage, 61, 166
Radiolaria, 28, 46, **48**, 49, 201
Radiolarian ooze, 49
Raphe, 43
Rathkea, 69, **72**, 75
Red tide, 49
Reference, 4, 15, 16, 67, 161, 190
Reproduction, 26, 29, 49, 57, 63, 97, 167, 171
Resorption, 187, 201
Respiration, 4, 27, 97, 129, 161, 197, 202
Rhabdonema, **40**, 43
Rhizosolenia, 29, **32**
Rhodomonas, 14
Ribbon worm, 85
Ring canal, 67
Rock crabs, 145
Rostrum, 132, 141, 145, 202
Rotifera, 93, **94**
Round worm, 93, **94**

Sabellariidae, **108**, 111
Saccoglossus, 183
Sagitta, **172**
Salinity, 3
Salpa, **191**
Salps, 136, 185, 190, **191**
Salt water mite, 111, **112**
Sand crabs, 141, 145
Sand dollars, 175, 180
Santa Cruz, 75, 171, 183
Sarsia, **72**

Scavengers, 56
Sceletonema, 14, **34**
Schistomysis, **131**
Schröderella, **36**
Sciaenidae, **193**
Scrippsia, **73**
Scyphistoma, 65, **66**, 202
Scyphozoa, 63, 65, **66**
Sea butterflies, 158
Sea cucumbers, 176, 180, 197
Sea hares, 161
Sea spiders, 111
Sea squirts, 185
Sea stars, 175, 176
Sea urchins, 57, 60, 175, 180
Segmentation, 104, 153
Segments, 57, 93, 99, 104, 109, 113, 120, 200, 201, 202
Sensory organs, 200
Septa, 63, 199, 201, 202
Sergestidae, 141
Sertularia, **77**, 78
Sessile, 46, 115, 185, 201, 202
Setae, 16, 56, 109, 115
Sexual, 29, 49, 57, 78, 200
Shore crabs, 145
Shrimp, 109, 129, 136, 139, 141, 145, 200
Silica, 29
Silk, 2, 9
Siphonophora, 12, 78, **79**, 83, 197, 198
Siphonostomatoida, **127**
Sipuncula, 59, 93, **95, 96**, 97
Skeleton shrimp, 136
Snails, 1, 155, 161
Somites, 129, 202
Specialization, 55, 113
Sperm, 57, 167, 175, 199
Sphaeronectes, **79**
Spicules, 63, 202
Spines, 4, 49, 115, 119, 145, 173, 199
Spionidae, 99, **102**, 111
Spiral cleavage, 166
Spirontocaris, **143**
Sponge, 63, 197, 201, 202
Spores, 199
Squid, 58, 163, 166
Standard net, 7, 8
Starfish, 175, 176
Statocyst, 67, 68, 129, 202
Staurocladia, **74**
Stephanopyxis, **32**
Stolon, 169, 202
Stomatopoda, **133**, 139
Streptotheca, **36**
Striatella, **40**
Strobilidium, **44**
Strombidinopsis, **44**
Strongylocentrotus, 57
Styela, 187
Stylomysis, **130**
Subumbrella, 67, 68, 75, 200, 202
Suckers, 109
Sugars, 26

220

Index

Sunlight, 3, 26
Swimmerets, 139
Syllidae, 99, **101**
Symbiotic, 111, 202
Synchaeta, **94**
Syncoryne, **72, 77**
Synedra, **41**
Syngnathidae, **196**

Tadpole Larva, 1, 59, 185, 187, **188**, 202
Tagmosis, 113, 202
Tail fan, 202
Tanaidacea, 129, 132, **133**
Taxonomy, 10, 27, 46, 53, 61, 113, 120
Teloblastic, 109, 202
Telotroch, 99, 169, 202
Telson, 113, 120, 132, 141, 202
Temperature, 3, 10, 11, 13, 14
Tentacles, 4, 56, 65, 67, 68, 69, 75, 78, 83, 104, 109, 167, 169, 200, 202
Terebellidae, **107**
Test, 14, 46, 202
Tetraclita, 115
Thalassinidea, 141
Thalassionema, **39**, 43
Thalassiosira, **34**
Thalassiothrix, **40**
Thaliacea, 185, 190, **191**
Thecosomata, 158
Themiste, 93, 97
Thoracic brood pouch, 129, 132
Thorax, 120, 129, 139, 197
Tides, 3, 10, 49, 121, 158
Tigriopus, 120, **126**
Tintinnid, 43, **45**, 46, 200
Tisbe, **126**
Tomales Bay, 75, 109, 166, 176
Tomopteris, 99, **101**
Tonicella, 155
Tornaria Larva, 59, 183, **186**, 202
Trachylina, 67, 68, 69, **70**
Trachymedusae, 67, 69, **70**
Transmitted light, 11
Transport, 11
Trematoda, 89, **91**
Triceratium, **31**
Trochophore, 59, 93, **95-96**, 97, 99, **100**, 104, 153, 183, 200, 201, 202
Trophic, 3, 202
Trunk, 93
Tubulanus, 89
Tunic, 185, 202
Tunicata, 1, 185, 187, 190, 200, 202
Turbellaria, 89, **90-91**, 200
Turbidity, 3, 27
Typhloscolex, **108**

Uca, 153
Ultraplankton, 2

Uniramia, 111
Upwelling, 27, 202
Urchins, 57, 58, 60, 175, 180, 181
Urechis, 97
Urochordata, 59, 185, **186**, 187, **188-190**, 190, **191**, 202
Uropods, 113, 132, 141, 202
Urosome, 120, 202

Valves, 29, 43, 115, 161, 167, 169, 173, 197, 198, 199, 202
Velella, 78, **79**
Veliger, 12, 56, 59, 155, **156, 157**, 158, **159, 160**, 161, **162, 164**, 202
Velum, 56, 65, 67, 68, 69, 155, 158, 161, 202
Vertebrata, 185, 192
Vertical migration, 10
Vibilia, **137**

Water bath, 11, 14
Western Pacific, 63
Wings, 158

Xanthidae, 153

Yolk, 27, 56, 99, 104, 163, 180, 192, 199, 202

Zanclea, **74**
Zoea, 59, 113, **140, 142**, 145, **146-149**, 200
Zooecium, 167, 169, 202
Zooid, 167, 169, 197, 199, 200, 202
Zoology, 55
Zooplankton, 2, 3, 4, 14, 25, 43, 55, 56, 63, 202
Zygote, 55, 57, 198, 199, 202